Charles Seale-Hayne Library
University of Plymouth
(01752) 588 588
LibraryandITenquiries@plymouth.ac.uk

FUNDAMENTALS OF GEOPHYSICAL FLUID DYNAMICS

Earth's atmosphere and oceans exhibit complex patterns of fluid motion over a vast range of space and time scales. On the planetary scale they combine to establish the climate in response to solar radiation that is inhomogeneously absorbed by the materials comprising air, water, and land. Spontaneous, energetic variability arises from instabilities in the planetary-scale circulations, appearing in many different forms such as waves, jets, vortices, boundary layers, and turbulence. Geophysical fluid dynamics (GFD) is the science of all these types of fluid motion.

This textbook is a concise and accessible introduction to GFD for intermediate to advanced students of the physics, chemistry, and/or biology of Earth's fluid environment. The book was developed from the author's many years of teaching a first-year graduate course at the Department of Atmospheric and Oceanic Sciences, University of California, Los Angeles. Readers are expected to be familiar with physics and mathematics at the level of general dynamics (mechanics) and partial differential equations.

JIM MCWILLIAMS is the Louis B. Slichter Professor of Earth Sciences in the Department of Atmospheric and Oceanic Sciences, Institute of Geophysics and Planetary Physics, University of California at Los Angeles. He is also a Senior Research Scientist at the National Center for Atmospheric Research, Boulder. He has undertaken research in the theory and computational modeling of Earth's ocean and atmosphere for the last 30 years. Professor McWilliams has taught a course on the fundamentals of geophysical fluid dynamics at UCLA for several decades. He is a Fellow of the American Geophysical Union, and a member of the National Academy of Sciences.

FUNDAMENTALS OF
GEOPHYSICAL FLUID DYNAMICS

JAMES C. McWILLIAMS

Department of Atmospheric and Oceanic Sciences
University of California, Los Angeles

CAMBRIDGE
UNIVERSITY PRESS

CAMBRIDGE UNIVERSITY PRESS
Cambridge, New York, Melbourne, Madrid, Cape Town, Singapore, São Paulo

Cambridge University Press
The Edinburgh Building, Cambridge CB2 2RU, UK

Published in the United States of America by Cambridge University Press, New York

www.cambridge.org
Information on this title: www.cambridge.org/9780521856379

First published 2006

Printed in the United Kingdom at the University Press, Cambridge

A catalog record for this publication is available from the British Library

ISBN-13 978-0-521-85637-9 hardback
ISBN-10 0-521-85637-X hardback

Contents

Colour plate section appears between pages 94 and 95

Preface

Earth's atmosphere and oceans exhibit complex patterns of fluid motion over a vast range of space and time scales. On the planetary scale they combine to establish the climate in response to solar radiation that is inhomogeneously absorbed by the materials comprising air, water, and land. Spontaneous, energetic variability arises from instabilities in the planetary-scale circulations, appearing in many different forms such as waves, jets, vortices, boundary layers, and turbulence. Geophysical fluid dynamics (GFD) is the science of all these types of fluid motion. It seeks to identify and analyze the essential dynamical processes that lie behind observed phenomena. As with any other theoretical science of complex nonlinear dynamics, mathematical analysis and computational modeling are essential research methodologies, and there is a continuing search for more powerful, accurate, and efficient techniques.

This book is an introduction to GFD for readers interested in doing research in the physics, chemistry, and/or biology of Earth's fluid environment. It is a product of teaching a first-year graduate course at the Department of Atmospheric and Oceanic Sciences, University of California, Los Angeles (UCLA) for many years. It is only an introduction to the subject; additional, more specialized GFD courses are required to fully prepare for practicing research in the subject. Nevertheless, to stimulate students' enthusiasm, the contents are a mixture of rudimentary mathematical analyses and somewhat complex dynamical outcomes. Students in this course are expected to be familiar with physics and mathematics at the level of general dynamics (mechanics) and partial differential equations. In the present graduate curriculum at UCLA, students are first exposed to one course on basic fluid dynamics and thermodynamics and another course on the principal phenomena of winds and currents and their underlying conceptual models. This background comprises the starting point for the book.

GFD is a mature subject, having had its adolescence in the middle of the last century. Consequently many meritorious books already exist. Most of them are specialized in their material, but several of the more general ones are usefully complementary to this book, e.g., Cushman-Roisin (1994), Gill (1982), Holton (2004), Pedlosky (1987), Salmon (1998), and Stern (1975).

Symbols

Symbols	Name	First usage
a	Earth's radius	Section 2.4
—[*]	boundary location	Eq. (4.61)
$\mathbf{a}$	initial position of a parcel	Eq. (2.1)
A	absolute momentum	Before Eq. (4.54)
—	wind gyre forcing amplitude	Eq. (6.62)
$\mathcal{A}$	horizontal area within $\mathcal{C}$	Eq. (3.17)
APE	available potential energy	Eq. (4.20)
b	pycnocline depth	Eq. (4.9)
—	buoyancy, $-g\rho/\rho_0$	Eq. (5.9)
B	topographic elevation	Eq. (4.1)
$\mathcal{B}$	Burger number	Eq. (4.105)
c, C	phase speed	Eqs. (3.94) and (5.57)
$\mathbf{c}_g$	wave group velocity	Eq. (4.34)
c_p	heat capacity (constant pressure)	Eq. (2.38)
$\mathbf{c}_p$	wave phase velocity	Eq. (4.33)
c_v	heat capacity (constant volume)	After Eq. (2.12)
C	circulation	Eq. (2.27)
C_s	speed of sound	Eq. (2.41)
$\mathcal{C}$	closed line	Eq. (2.27)
D	western boundary layer width	Eq. (6.62)
$\mathcal{D}$	isopycnal form stress	Eq. (5.87)
$\mathcal{D}_{bot}$	topographic form stress	Eq. (5.86)
D/Dt or D_t	substantial derivative	Eq. (2.3)
e	internal energy	Eq. (2.9)
$\hat{\mathbf{e}}$	unit vector	Eq. (2.57)

[*] the dash symbol denotes the same symbol with a different meaning.

Symbols	Name	First usage
E	volume-integrated total energy	Eq. (2.23)
—	Ekman number	Eq. (6.44)
$\mathbf{E}$	Eliassen–Palm flux	Eq. (5.103)
$\mathcal{E}$	local total energy density	Eq. (2.22)
Ens	enstrophy	Eq. (3.111)
f	Coriolis frequency	Eq. (2.89)
f_{h}	horizontal Coriolis frequency	Eq. (2.114)
$F(p)$	pressure coordinate	Eqs. (2.74) and (2.75)
$\mathbf{F}$	non-conservative force	Eq. (2.2)
F	boundary function	Eq. (2.13)
$\mathcal{F}$	$\hat{\mathbf{z}} \cdot \mathbf{\nabla} \times \mathbf{F}$	Eq. (3.24)
Fr	Froude number	Eq. (4.42)
g	gravitational acceleration	After Eq. (2.2)
g'	reduced gravity	Eq. (4.12)
g'_1	two-layer reduced gravity	Eq. (5.2)
$g'_{n+0.5}$	N-layer reduced gravity	Eq. (5.20)
G	pressure function	Eq. (2.85)
$G_m(n), G_m(z)$	modal transformation function	Eq. (5.29)
h	free-surface height	Eq. (2.17)
—	layer thickness	Eq. (4.1)
—	boundary-layer thickness	Section 6.1
h_{ek}	Ekman layer depth	After Eq. (6.44)
$h_{\mathrm{pycnocline}}$	depth of oceanic pycnocline	After Eq. (6.78)
h_*	sea-level with a rigid-lid approximation	Eq. (2.44)
—	turbulent Ekman layer thickness	Eq. (6.45)
H	oceanic depth	Section 2.2.3
—	atmospheric height	Eqs. (2.64)–(2.66)
—	vertical scale	Section 2.3.4
—	Hamiltonian function	Eq. (3.69)
H_{I}	oceanic interior thickness	Eq. (6.54)
i	$\sqrt{-1}$	After Eq. (2.70)
I	identity matrix	Eq. (5.42)
$\mathbf{I}$	identity vector	Eq. (5.41)
$\mathcal{I}$	vorticity angular momentum	Eq. (3.71)
J	Jacobian operator	Eq. (3.26)
k	x wavenumber	Eq. (3.32)
—	wavenumber vector magnitude, $|\mathbf{k}|$	After Eq. (3.113)

Symbols	Name	First usage
k_{E}	energy centroid wavenumber	Eq. (3.116)
$\mathbf{k}$	wavenumber vector	Eq. (3.112)
$\mathbf{k}_*$	dominant wavenumber component	Eq. (4.34)
K	wavenumber magnitude	Eq. (4.37)
—	von Karmen's constant	Eq. (6.49)
KE	kinetic energy	Eq. (3.2)
l, ℓ	y wavenumber	Eq. (3.32)
L	(horizontal) length scale	Before Eq. (2.5)
L_β	Rhines scale	Eq. (4.127)
—	inertial western boundary current width	Eq. (6.78)
L_x	zonal domain width	After Eq. (6.64)
L_y	meridional domain width	Section 5.3.1
L_τ	horizontal scale of wind stress	Section 5.3.1
m	azimuthal wavenumber	Eq. (3.76)
—	vertical mode number	Eq. (5.29)
M	Mach number	Eq. (2.41)
—	mass	Eq. (4.14)
n	vertical layer number	Eq. (5.18)
$\hat{\mathbf{n}}$	unit vector in normal direction	After Eq. (2.15)
$N(z)$	buoyancy frequency	Eq. (2.69)
N	number of vertical layers	Before Eq. (5.18)
$\mathcal{N}(z)$	buoyancy frequency	After Eq. (5.28)
$\mathbf{r}$	trajectory	Near Eq. (2.1)
p	pressure	Eq. (2.2)
P	oscillation period	After Eq. (2.70)
—	centrifugal pressure	Eq. (2.97)
—	potential vorticity matrix operator	Eq. (5.42)
PE	potential energy	Eq. (4.19)
$\mathcal{P}$	discriminant for baroclinic instabilty	Eq. (5.64)
q	specific humidity	After Eq. (2.12)
—	potential vorticity	Eqs. (3.28) and (4.24)
q_{QG}	quasigeostrophic potential vorticity	Eq. (4.113)
q_{E}	Ertel potential vorticity	Eq. (5.25)
q_{IPE}	isentropic potential vorticity for primitive equations	Eq. (5.24)
Q	potential vorticity	Eq. (4.56)
$\mathcal{Q}$	heating rate	Eq. (2.9)
$\tilde{\mathcal{Q}}$	potential heating rate	Eq. (2.52)

Symbols	Name	First usage
r	radial coordinate	Eq. (3.44)
—	damping rate	Eq. (5.104)
R	gas constant	Eq. (2.47)
—	deformation radius	Eq. (4.43)
R_e	external deformation radius	After Eq. (2.111)
R_m	deformation radius for mode m	Eq. (5.39)
Re	Reynolds number	Eq. (2.5)
Re_e	eddy Reynolds number	After Eq. (6.24)
Re_g	grid Reynolds number	Section 6.1.7
Ro	Rossby number	Eq. (2.102)
$\mathcal{R}$	horizontal Reynolds stress	After Eq. (3.98)
—	dispersion-to-advection ratio for Rossby waves	Eq. (4.124)
s	streamline coordinate	After Eq. (2.1)
—	instability growth rate	Eq. (3.88)
S	salinity	After Eq. (2.12)
—	strain rate	Fig. (2.3) and Eq. (3.51)
—	spectrum	Eq. (3.113)
—	stretching vorticity matrix operator	Eq. (5.46)
$\mathcal{S}$	non-conservative material source	Eq. (2.7)
—	material surface	Eq. (2.25)
$\mathcal{S}_f$	sign of f	Eq. (6.27)
t	time coordinate	Before Eq. (2.1)
t_d	spin-down time	Eq. (6.43)
T	time scale	After Eq. (2.5)
—	temperature	Eq. (2.11)
$\mathbf{T}$	depth-integrated horizontal column transport	Eq. (6.21)
$\mathbf{T}_{ek}$	Ekman layer horizontal column transport	Eq. (6.52)
$T_\perp$	horizontal volume transport	Eq. (6.74)
u	eastward velocity component	Before Eq. (2.2)
u_*	friction velocity	Eq. (6.45)
$\mathbf{u}$	vector velocity	Before Eq. (2.1)
$\mathbf{u}_g$	geostrophic horizontal velocity	Eq. (2.103)
$\mathbf{u}_a$	ageostrophic horizontal velocity	Before Eq. (4.112)
$\mathbf{u}^{st}$	Stokes drift	Eq. (4.95)
U	radial velocity	Eq. (3.45)

Symbols	Name	First usage
U	rotating-frame velocity	Eq. (2.93)
—	mean zonal velocity	Eq. (3.96)
—	depth-averaged zonal velocity	Eq. (6.59)
$\mathbf{U}^*$	eddy-induced velocity	After (5.98)
v	northward velocity component	Before Eq. (2.2)
V	(horizontal) velocity scale	Before Eq. (2.5)
—	rotating-frame velocity	Eq. (2.93)
—	azimuthal velocity	Eq. (3.45)
—	depth-averaged meridional velocity	Eq. (6.59)
V^*	northward eddy-induced velocity	Eq. (5.97)
$\mathcal{V}$	material volume	Eq. (2.25)
w	upward (vertical) velocity component	Before Eq. (2.2)
w_*	upward surface velocity with a rigid-lid approximation	Eq. (2.44)
w_{ek}	Ekman pumping velocity	Eq. (6.22)
w_{QG}	quasigeostrophic vertical velocity	Eq. (5.49)
W	vertical velocity scale	Section 2.3.4
W^*	upward eddy-induced velocity	Eq. (5.98)
x	eastward coordinate	Before Eq. (2.2)
$\mathbf{x}$	spatial position vector	Before Eq. (2.1)
$\hat{\mathbf{x}}$	unit eastward vector	Section 2.1.2
X	divergent velocity potential	Eq. (2.29)
$\mathbf{X}$	streamline	After Eq. (2.1)
$\mathbf{X} = (X, Y)$	rotating coordinate vector	Eq. (2.91)
—	streamfunction horizontal-centroid	Eq. (4.126)
$\mathcal{X}$	vorticity x-centroid	Eq. (3.71)
y	northward coordinate	Before Eq. (2.2)
$\hat{\mathbf{y}}$	unit northward vector	Section 2.1.2
$\mathcal{y}$	vorticity y-centroid	Eq. (3.71)
z	upward coordinate	Before Eq. (2.2)
z_o	roughness length	Eq. (6.49)
$\hat{\mathbf{z}}$	unit upward vector	Section 2.1.2
Z	geopotential height	After Eq. (2.38)
—	isentropic height	Eq. (5.24)
α	thermal expansion coefficient	Eq. (2.34)
—	point vortex index	Eq. (3.60)
β	haline contraction coefficient	Eq. (2.35)

Symbols	Name	First usage
β	Coriolis frequency gradient	Eq. (2.89)
—	point vortex index	Eq. (3.63)
γ	pressure expansion coefficient	Eq. (2.36)
—	gas constant ratio	After Eq. (2.51)
—	Reimann invariant	Eq. (4.85)
Γ	solution of characteristic equation	After Eq. (4.85)
δ	divergence	Eq. (2.24)
δ, Δ	incremental change	After Eq. (2.28), Fig. 2.3
$\delta_{p,q}$	discrete delta function	Eq. (3.108)
ϵ	wave steepness	Section 4.4
—	small expansion parameter	Eq. (4.106)
ϵ_{bot}	bottom damping coefficient	Before Eq. (5.80)
ζ, ζ^z	vertical vorticity	Eq. (3.5)
$\boldsymbol{\zeta}$	vector vorticity	Eq. (2.26)
η	entropy	Eq. (2.11)
—	interface height	Eq. (4.1)
θ	potential temperature	Eq. (2.51)
—	latitude	Eq. (2.87)
—	azimuthal coordinate	Eq. (3.44)
—	complex phase angle	Eq. (5.72)
Θ	wave phase function	After Eq. (4.92)
κ	diffusivity	After Eq. (2.8)
—	gas constant ratio	After Eq. (2.51)
λ	wavelength	After Eq. (4.33)
—	inverse Ekman layer depth	Eq. (6.29)
λ_0	phase constant	Eq. (2.120)
μ	chemical potential	Eq. (2.11)
—	$(KR)^{-2}$	Eq. (5.71)
ν	viscosity	After Eq. (2.2)
ν_e	eddy viscosity	Eqs. (3.102) and (6.23)
ν_h, ν_v	horizontal, vertical eddy viscosity	Eq. (5.80)
ξ	Lagrangian parcel displacement	Eq. (4.58)
—	characteristic coordinate	Eq. (4.86)
—	western boundary layer coordinate	Eq. (6.68)
ρ	density	Eq. (2.2)
ρ_{pot}	potential density	Eq. (2.51)
σ	instability growth rate	Eq. (3.79)
τ	material concentration	Eq. (2.7)

Symbols	Name	First usage
$\boldsymbol{\tau}_s$	surface stress	Before Eq. (5.80)
ϕ, Φ	geopotential function	Eqs. (2.38) and (2.80)
Φ	force potential	Eq. (2.2)
χ	divergent velocity potential	Eq. (2.29)
ψ	streamfunction	Eq. (2.29)
Ψ	transport streamfunction	Eq. (6.59)
ω	cross-isobaric velocity	Eq. (2.79)
—	oscillation frequency	Eq. (3.32)
$\Omega, \boldsymbol{\Omega}$	rotation rate, vector	Eq. (2.87)
$\boldsymbol{\Omega}_e$	Earth's rotation vector	Eq. (2.87)
∇	gradient operator	After Eq. (2.2)
∇_h	horizontal gradient operator	Eq. (2.31)
$\frac{\partial}{\partial z}$ or ∂_z	partial derivative with respect to, e.g., z	After Eq. (2.30)
$\bar{\cdot}$	averaging operator	Before Eq. (2.67)
$\langle \cdot \rangle$	zonal averaging operator	Eq. (3.97)
$\cdot^*$	complex conjugate	Eq. (3.66)
$\cdot'$	fluctuation operator	Eq. (3.72)
$\tilde{\cdot}$	modal coefficient	Eq. (5.29)

1

Purposes and value of geophysical fluid dynamics

In this book we will address a variety of topics that, taken together, comprise an introduction to *geophysical fluid dynamics* (GFD). The discussion is intended to be more about the concepts and methods of the subject rather than the specific formulae or observed phenomena. I hope they will be of both present interest and future utility to those who intend to work in Earth Sciences but do not expect to become specialists in the theory of dynamics, as well as to those who do have that expectation and for whom this is only a beginning.

Before starting I would like to make some preliminary remarks about the scope, purposes, and value of GFD.

The subject matter of GFD is motion in the fluid media on Earth and the distributions of material properties, such as mass, temperature, ozone, and plankton. (By common custom, planetary and astrophysical fluids are also included in GFD, since many of the scientific issues are similar, but it is awkward to use a more accurate title that explicitly includes all of these media. This book will not leave Earth.) So there is some chemistry, and even biology, in GFD, insofar as they influence the motion and evolution of the reactive materials. Nevertheless, for the most part GFD is a branch of physics that includes relevant aspects of dynamics, energy transfer by radiation, and the atomic and molecular processes associated with phase changes.

Yet GFD is by no means the entirety of ocean–atmosphere physics, much less its biogeochemistry. Within its subject-matter boundaries, GFD is distinguished by its purpose and its methodology. It is not principally concerned with establishing the facts about Earth's natural fluids, but rather with providing them a mathematical representation and an interpretation. These, in my opinion, are its proper purposes.

Beyond the knowledge provided by basic physics and chemistry, the facts about Earth's fluids are established in several ways:

- in the laboratory, where the constitutive relations, radiative properties, and chemical reactions are established, and where some analog simulations of natural phenomena are made;
- in the field, where measurements are made of the motion fields, radiation, and material property distributions;

1

- by theory, where the fundamental laws of fluid dynamics are well known, although – primarily because of their nonlinearity – only a small fraction of the interesting problems can actually be solved analytically; and
- on the computer, where relatively recent experience has demonstrated that simulations, based upon the fundamental relations established in the laboratory and theory as well as parameterizations of influential but unresolved processes, can approach the reality of nature as represented by the field measurements, but with much more complete information than measurements can provide.

In physical oceanography most of the pioneering laboratory work (e.g., the equation of state for seawater) has already been done, and so it is easy to take it for granted. This is also true for physical meteorology, but to a lesser degree: there remain important mysteries about the physical properties of water droplets, aerosols, and ice crystals, especially in clouds since it is difficult to simulate cloud conditions in the laboratory. For many decades and still today, the primary activity in physical oceanography is making measurements in the field. Field measurements are also a major part of meteorology, although computer modeling has long been a large part as well, initially through the impetus of numerical weather forecasting. Field measurements are, of course, quite important as the "measurable reality" of nature. But anyone who does them comes to appreciate how difficult it is to make good measurements of the atmosphere and ocean, in particular the difficulty in obtaining a broad space-time sampling that matches the phenomena. Computer simulations – the "virtual reality" of nature – are still primitive in various aspects of their scope and skillfulness, though they are steadily improving. There are successful examples of synoptic weather forecasting and design of engineering fluid devices (such as an airplane) to encourage us in this. One can also do analog simulations of geophysical fluid motions under idealized conditions in laboratory experiments. Some valuable information has been obtained in this way, but for many problems it is limited both by the usually excessive influence of viscosity, compared to nature, and by instrumental sampling limitations. Looking ahead it seems likely that computer simulations will more often be fruitful than laboratory simulations.

The facts that come from laboratory experiments, field measurements, and computer simulations are usually not simple in their information content. There is nothing simple about the equation of state for seawater, for example. As another example, a typical time series of velocity at a fixed location usually has a broadband spectrum with at most a few identifiable frequency lines that are rarely sharp (tides are an exception). Associated with this will be a generally decaying temporal lag correlation function, hence a finite time horizon of predictability. Furthermore, most geophysical time series are more appropriately called chaotic rather than deterministic, even though one can defend the use of governing dynamical equations that are deterministic in a mathematical sense but have the property of *sensitive dependence*, where any small differences amplify rapidly in time (Chapter 3). The complexity of geophysical motions is, in a generic way, a consequence of fluid turbulence. Even the tides, arising from spatially smooth, temporally periodic astronomical forcing, can be quite complex in their spatial response patterns. There is no reason to expect the relevant simulations to be appreciably simpler than the observations; indeed, their claim to credibility requires that they not be. An illustration of fluid dynamical complexity is the accompanying satellite image of sea surface temperature off the West Coast of the United States where coastal upwelling frequently occurs (Fig. 1.1). Figure 1.2 illustrates the comparable complexity of a computational simulation of this regime.

Arthur Eddington, the British astrophysicist, remarked, "Never trust an observation without a supporting theory." Facts about nature can be either important or trivial (i.e., generic or incidental) and can be grouped with other facts either aptly or misleadingly (i.e., causal or coincidental). Only a theory can tell you how to make these distinctions. For complex geophysical fluid motions, I think there is little hope of obtaining a fundamental theory that can be applied directly to most observations. Perhaps the Navier–Stokes equation (Chapter 2) is the only fundamental theory for fluid dynamics, albeit only in a highly implicit form. Since it cannot be solved in any general way, nor can it even be generally proven that unique, non-singular solutions exist, this theory is often opaque to any observational comparison except through some simulation that may be no easier to understand than the observations. Therefore, for geophysics I prefer a rephrasing of the remark to the more modest, "Never trust a fact, or a simulation, without a supporting interpretation."

It is the purpose of GFD to provide interpretations, and its methodology is idealization and abstraction, i.e., the removal of unnecessary geographic detail and contributing dynamical processes. Insofar as an observed or simulated fact can be identified as a phenomenon that, in turn, can be reproduced in the solution of a simple model, then the claim can be made (or, to be more cautious, the hypothesis advanced) that the essential nature of the phenomenon, including the essential ingredients for its occurrence, is understood. And this degree of understanding is possibly as good as can be hoped for, pending uncertain future insights. The proper practice of GFD, therefore, is to identify generic phenomena, and devise and solve simple models for them. The scientist who comes up with the simplest, relevant model wins the prize! Occam's Razor ("given two theories consistent with the known facts, prefer the one that is simpler or involves fewer assumptions") is an important criterion for judging GFD.

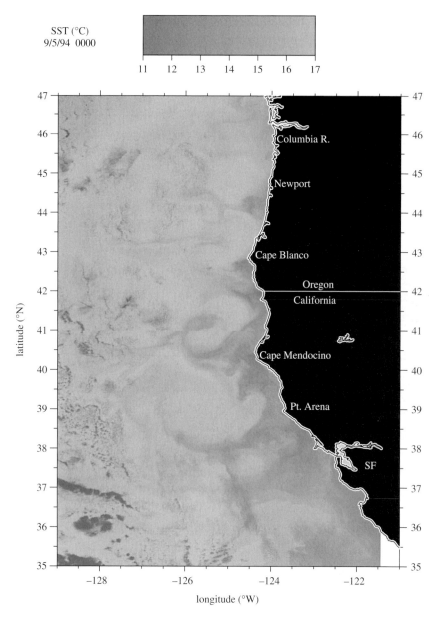

Fig. 1.1. Sea surface temperature (SST) off the US West Coast on 5 September 1994, measured with a satellite radiometer. The water near the coastline is much colder due to upwelling of cold sub-surface water. The upwelling is caused by an equatorward along-shore wind stress in association with a horizontally divergent, off-shore Ekman flow in the upper ocean (Chapter 6) as well as an along-shore surface geostrophic current (Chapter 2). The along-shore current is baroclinically unstable (Chapter 5) and generates mesoscale vortices (Chapter 3) and cold filaments advected away from the boundary, both with characteristic horizontal scales of 10–100 km. The light patches to the left are obscuring clouds. (Courtesy of Jack Barth and Ted Strub, Oregon State University.)

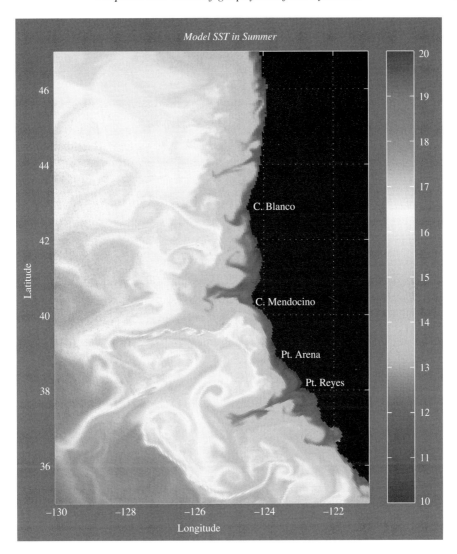

Fig. 1.2. Color plate 1.   Sea surface temperature off the US West Coast in late summer, from a numerical oceanic model. Note the general pattern similarity with Fig. 1.1 for cold upwelled water near the coastline, mesoscale vortices, and cold filaments advected away from the boundary. However, the measured and simulated patterns should not be expected to agree in their individual features because of the sensitive dependence of advective dynamics. (Marchesiello *et al.*, 2003.)

An objection might be raised that since computers will always be smaller than the universe, or even the atmosphere and the ocean, then any foreseeable simulated virtual reality can itself only be an abstraction and an idealization of nature, and thus no different, in principle, from a GFD model. While literally this is true, there is such an enormous and growing gap in complexity between the most accurate simulation models and simple GFD models of idealized situations that I believe this objection can be disregarded in practice. Nevertheless, the finite scope of geophysical simulation models must be conceded, and in doing so another important purpose for GFD must be recognized: to provide simple models for the effects of physically necessary, but computationally unresolved, processes in a simulation model. This is often called *parameterization*. The most common reason for parameterization is that something essential happens on a spatial or temporal scale smaller than the computational grid of the simulation model. Two examples of necessary parameterizations are (1) the *transport* (i.e., systematic spatial movement of material and dynamical properties by the flow) by turbulent eddies in a planetary boundary layer near the surface of the land or ocean and (2) the radiative energy transfer associated with cloud water droplets in the context of a global simulation model. Each of these micro-scale phenomena could be made simulation subjects in their own right, but not simultaneously with the planetary- or macro-scale *general circulation*, because together they would comprise too large a calculation for current or foreseeable computers. Micro-scale simulations can provide facts for GFD to interpret and summarily represent, specifically in the form of a useful parameterization.

Dynamical theory and its associated mathematics are a particular scientific practice that is not to everyone's taste, nor one for which every good scientist has a strong aptitude. Nevertheless, even for those who prefer working closer to the discovery and testing of facts about the ocean and atmosphere, it is important to learn at least some GFD since it provides one of the primary languages for communicating and judging the facts. Nature's facts are infinite in number. But which facts are the interesting ones? And how does one decide whether different putative facts are mutually consistent or not (and thus unlikely both to be true)? The answer usually is found in GFD.

Since this book is drawn from a course that lasts only three months, it helps to take some short cuts. One important short cut is to focus, where possible, on dynamical equations that have only zero (e.g., a fluid parcel), one, or two spatial dimensions, although nature has three. The lower-dimensional equations are more easily analyzed, and many of their solutions are strongly analogous to the solutions of three-dimensional dynamical equations that are more literally relevant to natural phenomena. Another short cut is to focus substantially on linear and/or steady solutions since they too are more easily analyzed, even though most oceanic and atmospheric behaviors are essentially transient and appreciably influenced by nonlinear dynamics (turbulence). In particular, pattern complexity and chaos (illustrated in Fig. 1.1 for a coastal sea surface temperature pattern) are widespread and essentially the result of nonlinearity in the governing equations. Nevertheless, the study of GFD properly starts with simpler reduced-dimensional, linear, and steady solutions that provide relevant, albeit incomplete, paradigms.

A list of symbols, exercises, and an index are included to help make this book a useful learning tool.

2

Fundamental dynamics

This chapter establishes, but does not fully derive, the basic equations of geophysical fluid dynamics and several of their most commonly used approximate forms, such as incompressible, Boussinesq, hydrostatic, and geostrophic equations. It also includes some particular solutions of these equations in highly idealized circumstances. Many more solutions will be examined in later chapters.

2.1 Fluid dynamics

2.1.1 Representations

For the most part the governing equations of fluid dynamics are partial differential equations in space ($\mathbf{x}$) and time (t). Any field (i.e., a property of the fluid), q, has an *Eulerian expression* as $q(\mathbf{x}, t)$. Bold face symbols denote vectors. Alternatively, any field also has an equivalent *Lagrangian expression* as $q(\mathbf{a}, t)$, where $\mathbf{a}$ is the $\mathbf{x}$ value at $t = 0$ of an infinitesimal fluid element (or *material parcel*) and $\mathbf{r}(\mathbf{a}, t)$ is its subsequent $\mathbf{x}$ value moving with the local fluid velocity, $\mathbf{u}$

$$\frac{d\mathbf{r}(t)}{dt} = \frac{\partial \mathbf{r}(\mathbf{a}, t)}{\partial t} = \mathbf{u}(\mathbf{x}, t)\Big|_{\mathbf{x}=\mathbf{r}}, \quad \mathbf{r}(\mathbf{a}, 0) = \mathbf{a}. \tag{2.1}$$

$\mathbf{r}$ is the *trajectory* of the parcel initially at $\mathbf{a}$ (Fig. 2.1). A line tangent to $\mathbf{u}$ everywhere at a fixed time, $t = t_0$, is a *streamline*, $\mathbf{X}(s, t_0)$, where s is the spatial coordinate along the streamline. Thus,

$$\frac{d\mathbf{X}}{ds} \times \mathbf{u} = 0.$$

If

$$\frac{d\mathbf{X}}{ds} = \mathbf{u},$$

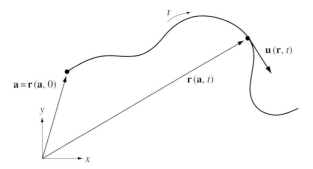

Fig. 2.1. The geometry of a trajectory, $\mathbf{r}(\mathbf{a}, t)$, projected onto the (x, y) plane. $\mathbf{a}$ is the position of the fluid parcel at time, $t = 0$, and the parcel moves along the trajectory with velocity, $\mathbf{u}(\mathbf{r}, t)(\mathbf{a})$, as indicated in (2.1).

then s has a normalization as a pseudo-time of movement along the streamline that would be equivalent to real time if the flow were stationary (i.e., $\partial_t \mathbf{u} = 0$). Alternatively, a *streakline* is the line traced in space of particles released continuously in time from a single point (which is experimentally much easier to determine by dye release and photography than a streamline). In a stationary flow streamlines, streaklines, and trajectories are all equivalent.

2.1.2 Governing equations

The starting point is the fundamental dynamical equations for a compressible fluid in a Cartesian coordinate frame – transformations can always be made to alternative frames such as a rotating spherical coordinate frame for planetary flows – with a general equation of state and variable material composition. For further discussion of basic fluid dynamics, refer to Batchelor (1967).

In GFD it is customary to associate the coordinate z with the vertical direction, parallel to the gravitational force and directed outward from Earth's center; x with the eastward direction; and y with the northward direction. It is also common usage to refer to the (x, y) directions as *zonal* and *meridional*, in association with longitude and latitude. The associated directional vectors with unit magnitude are denoted by $\hat{\mathbf{z}}$, $\hat{\mathbf{x}}$, and $\hat{\mathbf{y}}$, respectively, and the accompanying velocity components are by w, u, and v.

Momentum

A balance of acceleration and forces (i.e., Newton's law, $F = ma$, where F is force, m is mass, a is acceleration, and $m \times a$ is momentum) is expressed by the

following equation involving the time derivative of velocity (i.e., the acceler~
vector):

$$\frac{D\mathbf{u}}{Dt} = -\frac{1}{\rho}\nabla p + \nabla\Phi + \mathbf{F}.$$

This is referred to as the Navier–Stokes equation. Here $\mathbf{u}$ is the velocity [w
$\mathrm{m\,s^{-1}}$]; ρ is the density $[\mathrm{kg\,m^{-3}}]$; p is the pressure $[\mathrm{kg\,m^{-1}\,s^{-2}}$ or, equiv.
1 Pa (for pascal)]; Φ is the force potential $[\mathrm{m^2\,s^{-2}}]$ (e.g., for gravity, $\Phi = -gz$,
with $g = 9.81\,\mathrm{m\,s^{-2}}$); and $\mathbf{F}\,[\mathrm{m\,s^{-2}}]$ is all the *non-conservative forces* that do not
appear in Φ (e.g., molecular viscous diffusion with $\mathbf{F} = \nu\nabla^2\mathbf{u}$ and viscosity, ν).
∇ is the spatial gradient operator. The *substantial time derivative* is the acceleration
of a fluid parcel in a reference frame moving with the flow,

$$\frac{D}{Dt} = \frac{\partial}{\partial t} + \mathbf{u}\cdot\nabla. \tag{2.3}$$

The second term is called the advective operator, or more briefly *advection*;
it represents the movement of material with the fluid. (For notational compactness
we sometimes abbreviate these and other derivatives by D_t, ∂_t, etc.)

The Eulerian counterpart of the trajectory equation (2.1) is

$$\frac{D\mathbf{x}}{Dt} = \mathbf{u}, \tag{2.4}$$

which is a tautology given the definition (2.3). It means that the velocity is the
rate of change with time of the coordinate as it moves with the fluid.

The essential nonlinearity of fluid dynamics – the source of instability, chaos,
and turbulence – appears in the quadratic product of velocities that is the advection
of momentum. Advection also is a prevalent influence on the evolution of material
tracer distributions (in (2.7) below) that necessarily move with the flow. This leads
to three common statements about fluid dynamics, in general, and geophysical
fluid dynamics, in particular. The first statement is that the effect of advection
usually dominates over molecular diffusion. In a scale estimation analysis, if V is a
characteristic velocity scale and L is a characteristic length scale for flow variation,
then advective dominance is expressed as the largeness of the *Reynolds number*,

$$Re = \frac{VL}{\nu} \gg 1. \tag{2.5}$$

Since typical values for ν are $10^{-5}\,\mathrm{m^2\,s^{-1}}$ (air) and $10^{-6}\,\mathrm{m^2\,s^{-1}}$ (seawater), then
even a modest velocity difference of $V = 1\,\mathrm{m\,s^{-1}}$ (air) or $0.1\,\mathrm{m\,s^{-1}}$ (seawater)
over a distance of $L = 100\,\mathrm{m}$, has $Re = 10^7$, and even larger Re values occur for

stronger flows on larger scales. The second, related statement is that, in such a situation, the typical time scale of evolution is at an *advective time*, $T = L/V$, which is the passage time for some material pattern to be carried past a fixed $\mathbf{x}$ point. This advective dominance is because a diffusive evolution time, $T = L^2/\nu$, is much longer, hence relatively ineffective on the shorter advective time. The ratio of these diffusive and advective times is $Re \gg 1$. The third statement is that almost all flows are unstable and full of fluctuations with an advective time scale once Re is above a critical value of $\mathcal{O}(10–100)$; this contrasts with stable, smooth *laminar flow* without fluctuations when Re is smaller.

Mass

A fluid by definition is comprised of continuous material, without any ruptures in space. It can have no interior sources or sinks of mass for the primary composition of the fluid, i.e., air in the atmosphere and water in the ocean (as opposed to the minor constituent components, the material tracers, whose fractional proportions can vary greatly). This is expressed as a mass-conservation balance related to the fluid density, ρ, associated with the primary composition:

$$\frac{\partial \rho}{\partial t} + \mathbf{V} \cdot (\rho \mathbf{u}) = 0,$$

or

$$\frac{D\rho}{Dt} = -\rho \mathbf{V} \cdot \mathbf{u}.$$

(2.6)

This is also called the *continuity equation*.

Material tracer

For any gaseous (air) or dissolved (water) material concentration, τ [mass fraction relative to the primary fluid component, or *mixing ratio*], other than the primary fluid composition, the concentration evolution equation is

$$\frac{\partial (\rho \tau)}{\partial t} + \mathbf{V} \cdot (\rho \tau \mathbf{u}) = \rho \mathcal{S}^{(\tau)},$$

(2.7)

or, using (2.6),

$$\frac{D\tau}{Dt} = \mathcal{S}^{(\tau)},$$

(2.8)

where $\mathcal{S}^{(\tau)}[\mathrm{s}^{-1}]$ is all the non-conservative sources and sinks of τ (e.g., chemical reaction rates or material diffusion, with $\mathcal{S} = \kappa \mathbf{V}^2 \tau$ and a diffusivity, κ). For Earth's air and water, κ is usually of the order of ν (i.e., their ratio, the *Prandtl number*, ν/κ, is $\mathcal{O}(1)$), so advection usually dominates molecular diffusion in (2.8) for the same reason that Re is usually large. When $\mathcal{S}^{(\tau)}$ is negligible, then the movement of the material traces the flow, hence the terminology for τ.

Internal energy

For the internal energy, e [m^2 s^{-2}],

$$\rho\frac{De}{Dt} = -p\mathbf{\nabla}\cdot\mathbf{u} + \rho\mathcal{Q},\tag{2.9}$$

or, again using (2.6),

$$\frac{De}{Dt} = -p\frac{D}{Dt}\left(\frac{1}{\rho}\right) + \mathcal{Q},\tag{2.10}$$

where $\mathcal{Q}$ [m^2 s^{-3}] is the heating rate per unit mass. This equation is sometimes referred to as the first law of thermodynamics: the energy of the universe is constant, and the internal energy of a fluid sub-system (i.e., here the internal energy, e) only changes through work done by compression (i.e., pressure times the volume change) or by heating (i.e., $\mathcal{Q}$) by dissipation of mechanical energy into heat, chemical reaction, electromagnetic radiation, phase change, or exchange with the rest of the universe. (Of course, this is not the most fundamental statement of energy conservation in the laws of physics, but it is general enough for most fluid dynamical purposes when combined with additional equations for kinetic and potential energy that are derived from the other governing equations; e.g., Sections 2.1.4, 3.1, and 4.1.1.)

Entropy

For the fluid entropy, η [m^2 s^{-2} K^{-1}],

$$T\frac{D\eta}{Dt} = \mathcal{Q} - \sum_k \mu_k \mathcal{S}^{(\tau_k)}.\tag{2.11}$$

Here T [K] is the temperature and μ_k [m^2 s^{-2}] is the chemical potential for the tracer species, τ_k. This equation is related to the second law of thermodynamics: the entropy of the universe can only increase, and the entropy of the fluid part, η, changes only through its heat and material exchanges with the rest of the universe.

Equation of state

In addition to the preceding equations expressing spatial flux and source/sink effects on the evolution of the fluid fields, there is a required thermodynamic statement about the density of the fluid in the form of

$$\rho = \rho(T, p, \tau_k),\tag{2.12}$$

where the right-hand side arguments are referred to as the state variables. The equation of state differs for different types of fluids. For the ocean the important tracer state variable is the salinity, S [practical salinity unit (PSU) = parts per thousand (ppt) = $10^3 \times$ mass fraction]; for the atmosphere it is the specific humidity, q [mass fraction]. Typical values for ρ are $\mathcal{O}(10^3)$ kg m^{-3} (ocean) and $\mathcal{O}(1)$ kg m^{-3} near Earth's surface and decreasing upward to space (air).

Other thermodynamic relations are also needed between e, η, μ_k, and the state variables; e.g., $e = c_v T$ for an ideal gas (Section 2.3.1), where c_v is the heat capacity of the fluid at constant volume. Thermodynamic relations are also needed for other material properties such as ν, κ, and the speed of sound, C_s (Section 2.3.1). In general, the thermodynamics of composite fluids is complicated and subtle, and in this book we will only consider some of its simpler forms. Fuller discussions of the oceanic and atmospheric thermodynamics are in Fofonoff (1962), Bohren and Albrecht (1998), and Gill (Chapters 3–4, 1982).

An important distinction in GFD is between *conservative* and *non-conservative* motions. The former refer to governing equation sets that imply that both the volume-integrated total energy and the material concentrations on every fluid parcel do not change in time. Thus, the sources of non-conservation include external forces other than gravity in $\mathbf{F}$, heating other than by compression in $\mathcal{Q}$, material sources in $\mathcal{S}$, and molecular diffusion with ν or κ in $\mathbf{F}$, $\mathcal{Q}$, or $\mathcal{S}$.

2.1.3 Boundary and initial conditions

When a fluid has a well defined boundary across which there is no relative motion, then the boundary must be a material surface that retains its parcels. This is equivalent to saying that the flow must either turn to be parallel to the boundary, and/or it must move perpendicular to the boundary at exactly the same speed as the boundary itself is moving in that direction. The mathematical expression for this is called the *kinematic boundary condition*. If

$$F(\mathbf{x}, t) = 0 \tag{2.13}$$

is a mathematical definition of the boundary location, then the kinematic boundary condition is

$$\frac{DF}{Dt} = 0 \quad \text{at} \quad F = 0. \tag{2.14}$$

Some particular situations are the following.

(a) A stationary boundary at $\mathbf{x}_0$:

$$\mathbf{u} \cdot \hat{\mathbf{n}} = 0 \quad \text{at} \quad \hat{\mathbf{n}} \cdot [\mathbf{x} - \mathbf{x}_0] = 0; \tag{2.15}$$

e.g., for $F = x - x_0$, the outward unit normal vector is $\hat{\mathbf{n}} = \hat{\mathbf{x}}$ and $u = 0$ at $x = x_0$.

(b) A moving boundary at $\mathbf{x}_0(t)$:

$$\mathbf{u} \cdot \hat{\mathbf{n}} = \frac{d\mathbf{x}_0}{dt} \cdot \hat{\mathbf{n}} \quad \text{at} \quad \hat{\mathbf{n}} \cdot [\mathbf{x} - \mathbf{x}_0] = 0. \tag{2.16}$$

(c) A *free surface* at $z = h(x, y, t)$ (i.e., an impermeable, moving interface between two fluid regions, such as the top surface of a water layer with air above), with $F = z - h$:

$$w = \frac{Dh}{Dt} \quad \text{at} \quad z = h. \tag{2.17}$$

There are two other common types of boundary conditions, a *continuity boundary condition* (e.g., the continuity of pressure across the air–sea interface) and a *flux boundary condition* (e.g., the flux of water into the atmosphere due to evaporation minus precipitation at the sea or land surface). The combination of boundary conditions that is appropriate for a given situation depends mathematically upon which partial differential equation system is being solved (i.e., to assure well-posedness of the boundary-value problem) and physically upon which external influences are being conveyed through the boundary.

Furthermore, *initial conditions* are also required for partial differential equations that contain time derivatives. Exactly how many fluid fields must have their initial distributions specified again depends upon which dynamical system is being solved. A typical situation requires initial conditions for velocity, temperature, density, and all material tracers.

The practice of GFD is full of different approximations, where apt, and some types of approximations change the mathematical character of the governing equation set and its requirements for well-posedness (e.g., Section 4.6).

2.1.4 Energy conservation

The principle of energy conservation is a basic law of physics, but in the context of fluid dynamics it is derived from the governing equations and boundary conditions (Sections 2.1.2–2.1.3) rather than independently specified.

The derivation is straightforward but lengthy. For definiteness (and sufficient for most purposes in GFD), we assume that the force potential is entirely gravitational, $\Phi = -gz$, or equivalently that any other contributions to $\nabla \Phi$ are absorbed into $\mathbf{F}$. Multiplying the momentum equation (2.2) by $\rho \mathbf{u}$ gives

$$\rho \frac{\partial}{\partial t} \left(\frac{1}{2} \mathbf{u}^2 \right) = -\mathbf{u} \cdot \nabla p - g\rho w - \rho \mathbf{u} \cdot \nabla \left(\frac{1}{2} \mathbf{u}^2 \right) + \rho \mathbf{u} \cdot \mathbf{F},$$

after making use of $w = D_t z$ from (2.4). Multiplying the mass equation (2.6) by $\mathbf{u}^2/2$ gives

$$\left(\frac{1}{2}\mathbf{u}^2\right)\frac{\partial \rho}{\partial t} = -\frac{1}{2}\mathbf{u}^2\,\mathbf{\nabla}(\rho \mathbf{u}).$$

The sum of these equations is

$$\frac{\partial}{\partial t}\left(\frac{1}{2}\rho \mathbf{u}^2\right) = p\mathbf{\nabla}\cdot\mathbf{u} - g\rho w - \mathbf{\nabla}\cdot\left(\mathbf{u}\left[p + \frac{1}{2}\rho \mathbf{u}^2\right]\right) + \rho \mathbf{u}\cdot\mathbf{F}. \qquad (2.18)$$

It expresses how the local *kinetic energy density*, $\rho \mathbf{u}^2/2$, changes as the flow evolves. (Energy is the spatial integral of energy density.) To obtain a principle for total energy density, $\mathcal{E}$, two other local conservation laws are derived to accompany (2.18). One comes from multiplying the mass equation (2.6) by gz, viz.,

$$\frac{\partial}{\partial t}(gz\rho) = g\rho w - \mathbf{\nabla}\cdot(\mathbf{u}[gz\rho]). \qquad (2.19)$$

This says how $gz\rho$, the local *potential energy density*, changes. Note that the first right-hand side term is equal and opposite to the first right-hand side term in (2.18); $g\rho w$ is therefore referred to as the local energy conversion rate between kinetic and potential energies. The second accompanying relation comes from (2.6) and (2.9) and has the form,

$$\frac{\partial}{\partial t}(\rho e) = -p\mathbf{\nabla}\cdot\mathbf{u} - \mathbf{\nabla}\cdot(\mathbf{u}[\rho e]) + \rho \mathcal{Q}. \qquad (2.20)$$

This expresses the evolution of local *internal energy density*, ρe. Its first right-hand side term is the conversion rate of kinetic energy to internal energy, $-p\mathbf{\nabla}\cdot\mathbf{u}$, associated with the work done by compression, as discussed following (2.10).

The sum of (2.18)–(2.20) yields the local energy conservation relation:

$$\frac{\partial \mathcal{E}}{\partial t} = -\mathbf{\nabla}\cdot(\mathbf{u}\,[p + \mathcal{E}]) + \rho\,(\mathbf{u}\cdot\mathbf{F} + \mathcal{Q}), \qquad (2.21)$$

where the total energy density is defined as the sum of the kinetic, potential, and internal components,

$$\mathcal{E} = \frac{1}{2}\rho \mathbf{u}^2 + gz\rho + \rho e. \qquad (2.22)$$

All of the conversion terms have canceled each other in (2.21). The local energy density changes either due to spatial transport (the first right-hand side group, comprised of pressure and energy flux divergence) or due to non-conservative force and heating. The energy transport term acts to move the energy from one location to another. It vanishes in a spatial integral except for whatever boundary energy fluxes there are because of the following calculus relation for any vector field, $\mathbf{A}$:

$$\iiint_{\mathcal{V}} d\,\text{vol}\ \nabla \cdot \mathbf{A} = \iint_{\mathcal{S}} d\,\text{area}\ \mathbf{A} \cdot \hat{\mathbf{n}},$$

where $\mathcal{V}$ is the fluid volume, $\mathcal{S}$ is its enclosing surface, and $\hat{\mathbf{n}}$ a locally outward normal vector on $\mathcal{S}$ with unit magnitude. Since energy transport often is a very efficient process, usually the most useful energy principle is a volume integrated one, where the total energy,

$$E = \iiint_{\mathcal{V}} d\,\text{vol}\ \mathcal{E}, \tag{2.23}$$

is conserved except for the boundary fluxes (i.e., exchange with the rest of the universe) or interior non-conservative terms such as viscous dissipation and absorption or emission of electromagnetic radiation.

Energy conservation is linked to material tracer conservation (2.7) through the definition of e and the equation of state (2.12). The latter relations will be addressed in specific approximations (e.g., Sections 2.2 and 2.3).

2.1.5 Divergence, vorticity, and strain rate

The velocity field, $\mathbf{u}$, is of such central importance to fluid dynamics that it is frequently considered from several different perspectives, including its spatial derivatives (below) and spatial integrals (Section 2.2.1).

The spatial gradient of velocity, $\nabla \mathbf{u}$, can be partitioned into several components with distinctively different roles in fluid dynamics.

Divergence

The *divergence*,

$$\delta = \nabla \cdot \mathbf{u} = \frac{\partial u}{\partial x} + \frac{\partial v}{\partial y} + \frac{\partial w}{\partial z}, \tag{2.24}$$

is the rate of volume change for a material parcel (moving with the flow). This is shown by applying Green's integral relation to the rate of change of a finite volume, $\mathcal{V}$, contained within a closed surface, $\mathcal{S}$, moving with the fluid:

$$\frac{d\mathcal{V}}{dt} = \iint_{\mathcal{S}} d\,\text{area}\ \mathbf{u} \cdot \hat{\mathbf{n}}$$

$$= \iiint_{\mathcal{V}} d\,\text{vol}\ \nabla \cdot \mathbf{u} = \iiint_{\mathcal{V}} d\,\text{vol}\ \delta. \tag{2.25}$$

$\hat{\mathbf{n}}$ is a locally outward unit normal vector, and d area and d vol are the infinitesimal local area and volume elements (Fig. 2.2a).

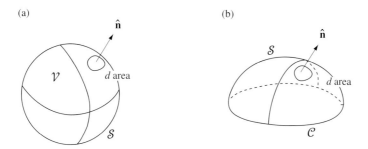

(a)   (b)

Fig. 2.2. (a) Volume element, $\mathcal{V}$, and its surface, $\mathcal{S}$, that are used in determining the relation between divergence and volume change following the flow (Green's integral relation). (b) Closed curve, $\mathcal{C}$, and connected surface, $\mathcal{S}$, that are used in determining the relation between vorticity and circulation (Stokes' integral relation).

Vorticity and circulation

The *vorticity* is defined by

$$\boldsymbol{\zeta} = \mathbf{\nabla} \times \mathbf{u}$$

$$= \hat{\mathbf{x}} \left(\frac{\partial w}{\partial y} - \frac{\partial v}{\partial z} \right) + \hat{\mathbf{y}} \left(\frac{\partial u}{\partial z} - \frac{\partial w}{\partial x} \right) + \hat{\mathbf{z}} \left(\frac{\partial v}{\partial x} - \frac{\partial u}{\partial y} \right). \qquad (2.26)$$

It expresses the local whirling rate of the fluid with both a magnitude and a spatial orientation. Its magnitude is equal to twice the angular rotation frequency of the swirling flow component around an axis parallel to its direction. A related quantity is the *circulation*, C, defined as the integral of the tangential component of velocity around a closed line $\mathcal{C}$. By Stokes' integral relation, it is equal to the area integral of the normal projection of the vorticity through any surface $\mathcal{S}$ that ends in $\mathcal{C}$ (Fig. 2.2b):

$$C = \int_{\mathcal{C}} \mathbf{u} \cdot \mathbf{dx} = \iint_{\mathcal{S}} d \text{ area } \boldsymbol{\zeta} \cdot \hat{\mathbf{n}}. \qquad (2.27)$$

Strain rate

The *velocity-gradient tensor*, $\mathbf{\nabla u}$, has nine components in three-dimensional space, 3D (or four in 2D). δ is one linear combination of these components (i.e., the trace of the tensor) and accounts for one component. ζ accounts for another three components (one in 2D). The remaining five linearly independent components (two in 2D) are called the strain rate, which has both three magnitudes and the spatial orientation of two angles (one and one, respectively, in 2D). The strain rate acts through the advective operator to deform the shape of a parcel as it moves, separately from its volume change (due to divergence) or rotation (due to

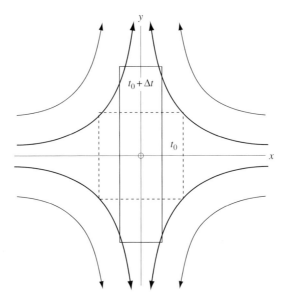

Fig. 2.3. The deformation of a material parcel in a plane strain flow defined by the streamfunction and velocity components, $\psi = \frac{1}{2}S_0 xy$, $u = -\partial_y\psi = -\frac{1}{2}S_0 x$, and $v = \partial_x\psi = \frac{1}{2}S_0 y$ (cf., (2.29)), with $\partial_x u - \partial_y v = S_0$ the spatially uniform strain rate. The heavy solid lines are isolines of ψ with arrows indicating the flow direction. The associated vorticity is $\zeta = 0$. The dashed square indicates a parcel boundary at $t = t_0$ and the solid rectangle indicates the same boundary at some later time, $t = t_0 + \Delta t$. The parcel is deformed by squeezing it in x and extruding it in y, while preserving the parcel area since the flow is non-divergent, $\delta = 0$.

vorticity). For example, in a horizontal plane the strain rate deforms a material square into a rectangle in a 2D uniform strain flow when the polygon sides are oriented perpendicular to the distant inflow and outflow directions (Fig. 2.3). (See Batchelor (Section 2.3, 1967) for mathematical details.)

2.2 Oceanic approximations

Almost all theoretical and numerical computations in GFD are made with governing equations that are simplifications of (2.2)–(2.12). Discussed in this section are some of the commonly used simplifications for the ocean, although some others that are equally relevant to the ocean (e.g., a stratified resting state or sound waves) are presented in the next section on atmospheric approximations. From a GFD perspective, oceanic and atmospheric dynamics have more similarities than differences, and often it is only a choice of convenience which medium is used to illustrate a particular phenomenon or principle.

2.2.1 Mass and density

Incompressibility

A simplification of the mass-conservation relation (2.6) can be made based on the smallness of variations in density:

$$\frac{1}{\rho}\frac{D\rho}{Dt} = -\mathbf{V}\cdot\mathbf{u} \ll \left|\frac{\partial u}{\partial x}\right|, \left|\frac{\partial v}{\partial y}\right|, \left|\frac{\partial w}{\partial z}\right|$$

$$\Rightarrow \mathbf{V}\cdot\mathbf{u} \approx 0 \quad \text{if} \quad \frac{\delta\rho}{\rho} \ll 1. \tag{2.28}$$

In this incompressible approximation, the divergence is zero, and material parcels preserve their infinitesimal volume, as well as their mass, following the flow (cf., (2.25)). In this equation the prefix δ means the change in the indicated quantity (here ρ). The two relations in the second line of (2.28) are essentially equivalent based on the following scale estimates for characteristic magnitudes of the relevant entities: $\mathbf{u} \sim V$, $\mathbf{V}^{-1} \sim L$, and $T \sim L/V$ (i.e., an advective time scale). Thus,

$$\frac{1}{\rho}\frac{D\rho}{Dt} \sim \frac{V}{L}\frac{\delta\rho}{\rho} \ll \frac{V}{L}.$$

For the ocean, typically $\delta\rho/\rho = \mathcal{O}(10^{-3})$, so (2.28) is quite an accurate approximation.

Velocity potential functions

The three directional components of an incompressible vector velocity field can be represented, more concisely and without any loss of generality, as gradients of two scalar potentials. This is called a *Helmholtz decomposition*. Since the vertical direction is distinguished by its alignment with both gravity and the principal rotation axis, the form of the decomposition most often used in GFD is

$$u = -\frac{\partial\psi}{\partial y} - \frac{\partial^2 X}{\partial x\partial z} = -\frac{\partial\psi}{\partial y} + \frac{\partial\chi}{\partial x}$$

$$v = \frac{\partial\psi}{\partial x} - \frac{\partial^2 X}{\partial y\partial z} = \frac{\partial\psi}{\partial x} + \frac{\partial\chi}{\partial y}$$

$$w = \frac{\partial^2 X}{\partial x^2} + \frac{\partial^2 X}{\partial y^2} = \nabla_h^2 X, \tag{2.29}$$

where $\mathbf{V}_h$ is the 2D (horizontal) gradient operator. This guarantees $\mathbf{V}\cdot\mathbf{u} = 0$ for any ψ and X. ψ is called the *streamfunction*. It is associated with the vertical component of vorticity,

$$\hat{\mathbf{z}}\cdot\mathbf{V}\times\mathbf{u} = \zeta^{(z)} = \nabla_h^2\psi, \tag{2.30}$$

(a)   (b)

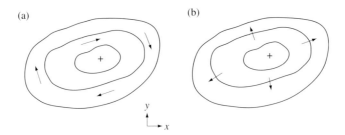

Fig. 2.4. Horizontal flow patterns in relation to isolines of (a) streamfunction, $\psi(x, y)$, and (b) divergent velocity potential, $\chi(x, y)$. The flows are along and across the isolines, respectively. Flow swirls clockwise around a positive ψ extremum and away from a positive χ extremum.

while X is not. Thus, ψ represents a component of horizontal motion along its isolines in a horizontal plane at a speed equal to its horizontal gradient, and the direction of this flow is clockwise about a positive ψ extremum (Fig. 2.4a). X (or its related quantity, $\chi = -\partial_z X$, where ∂_z is a compact notation for the partial derivative with respect to z) is often called the *divergent potential*. It is associated with the horizontal component of the velocity divergence,

$$\mathbf{V}_h \cdot \mathbf{u}_h = \frac{\partial u}{\partial x} + \frac{\partial v}{\partial y} = \delta_h = \nabla_h^2 \chi, \tag{2.31}$$

and the vertical motions required by 3D incompressibility, while ψ is not. Thus, isolines of χ in a horizontal plane have a horizontal flow across them at a speed equal to the horizontal gradient, and the direction of the flow is inward toward a positive χ extremum that usually has an accompanying negative δ_h extremum (e.g., think of $\sin x$ and $\nabla_h^2 \sin x = -\sin x$; Fig. 2.4b). Since

$$\frac{\partial w}{\partial z} = -\delta_h = -\nabla_h^2 \chi, \tag{2.32}$$

the two divergent potentials, X and χ, are linearly related to the vertical velocity, while ψ is not. When the χ pattern indicates that the flow is coming together in a horizontal plane (i.e., converging, with $\nabla_h^2 \chi < 0$), then there must be a corresponding vertical gradient in the normal flow across the plane in order to conserve mass and volume incompressibly.

Linearized equation of state

The equation of state for seawater, $\rho(T, S, p)$, is known only by empirical evaluation, usually in the form of a polynomial expansion series in powers of the departures of the state variables from a specified reference state. However, it is sometimes more simply approximated as

$$\rho = \rho_0 \left[1 - \alpha(T - T_0) + \beta(S - S_0) \right]. \tag{2.33}$$

Here the linearization is made for fluctuations around a reference state of (ρ_0, T_0, S_0) (and implicitly a reference pressure, p_0; alternatively one might replace T with the potential temperature (θ; Section 2.3.1) and make p nearly irrelevant). Typical oceanic values for this reference state are $(10^3 \, \text{kg m}^{-3}, 283 \, \text{K} (10°\text{C}), 35 \, \text{ppt})$. In (2.33),

$$\alpha = -\frac{1}{\rho} \frac{\partial \rho}{\partial T} \tag{2.34}$$

is the *thermal expansion coefficient* for seawater and has a typical value of $2 \times 10^{-4} \, \text{K}^{-1}$, although this varies substantially with T in the full equation of state; and

$$\beta = +\frac{1}{\rho} \frac{\partial \rho}{\partial S} \tag{2.35}$$

is the *haline contraction coefficient* for seawater, with a typical value of $8 \times 10^{-4} \, \text{ppt}^{-1}$. In (2.34) and (2.35) the partial derivatives are made with the other state variables held constant. Sometimes (2.33) is referred to as the *Boussinesq equation of state*. From the values above, either a $\delta T \approx 5$ K or a $S \approx 1$ ppt implies a $\delta\rho/\rho \approx 10^{-3}$ (cf., Fig. 2.7).

Linearization is a type of approximation that is widely used in GFD. It is generally justifiable when the departures around the reference state are small in amplitude, e.g., as in a Taylor series expansion for a function, $q(x)$, in the neighborhood of $x = x_0$:

$$q(x) = q(x_0) + (x - x_0)\frac{dq}{dx}(x_0) + \frac{1}{2}(x - x_0)^2 \frac{d^2 q}{dx^2}(x_0) + \cdots .$$

For the true oceanic equation of state, (2.33) is only the start of a Taylor series expansion in the variations of (T, S, p) around their reference state values. Viewed globally, α and β show significant variations over the range of observed conditions (i.e., with the local mean conditions taken as the reference state). Also, the actual compression of seawater,

$$\gamma \, \delta p = \frac{1}{\rho} \frac{\partial \rho}{\partial p} \delta p, \tag{2.36}$$

is of the same order as $\alpha \, \delta T$ and $\beta \, \delta S$ in the preceding paragraph, when

$$\delta p \approx \rho_0 g \, \delta z \tag{2.37}$$

and $\delta z \approx 1$ km. This is a *hydrostatic* estimate in which the pressure at a depth δz is equal to the weight of the fluid above it. The compressibility effect on ρ may not often be dynamically important since few parcels move 1 km or more vertically in the ocean except over very long periods of time, primarily because of the large amount of work that must be done converting fluid kinetic energy to overcome the potential energy barrier associated with stable density stratifi-cation (cf., Section 2.3.2). Thus, (2.33) is more a deliberate simplification than a universally accurate approximation. It is to be used in situations when either the spatial extent of the domain is not so large as to involve significant changes in the expansion coefficients or when the qualitative behavior of the flow is not controlled by the quantitative details of the equation of state. (This may only be provable a posteriori by trying the calculation both ways.) However, there are situations when even the qualitative behavior requires a more accurate equation of state than (2.33). For example, at very low temperatures a *thermobaric insta-bility* can occur when a parcel in an otherwise stably stratified profile (i.e., with monotonically varying $\rho(z)$) moves adiabatically and changes its p enough to yield an anomalous ρ compared to its new environment, which induces a further vertical acceleration as a *gravitational instability* (cf., Section 2.3.3). Furthermore, a *cabelling instability* can occur if the mixing of two parcels of seawater with the same ρ, but different T and S yields a parcel with the average values for T and S but a different value for ρ – again inducing a gravitational instability with respect to the unmixed environment. The general form for $\rho(T, S, p)$ is sufficiently nonlinear that such odd behaviors sometimes occur.

2.2.2 Momentum

With or without the use of (2.33), the same rationale behind (2.28) can be used to replace ρ by ρ_0 everywhere except in gravitational force and the equation of state. The result is an approximate equation set for the ocean that is often referred to as the incompressible *Boussinesq equations*. In an oceanic context that includes salinity variations, they can be written as

$$\frac{D\mathbf{u}}{Dt} = -\nabla\phi - g\frac{\rho}{\rho_0}\hat{\mathbf{z}} + \mathbf{F},$$

$$\nabla \cdot \mathbf{u} = 0,$$

$$\frac{DS}{Dt} = \mathcal{S},$$

$$c_{\mathrm{p}}\frac{DT}{Dt} = \mathcal{Q}. \tag{2.38}$$

(Note: they are commonly rewritten in a rotating coordinate frame that adds the Coriolis force, $-2\mathbf{\Omega} \times \mathbf{u}$, to the right-hand side of the momentum equation (Section 2.4).) Here $\phi = p/\rho_0$ [m^2 s^{-2}] is called the *geopotential function* (NB, the related quantity, $Z = \phi/g$ [m], is called the *geopotential height*), and $c_{\mathrm{p}} \approx 4 \times 10^3$ m^2 s^{-2} K^{-1} is the oceanic heat capacity at constant pressure. The salinity equation is a particular case of the tracer equation (2.8), and the temperature equation is a simple form of the internal energy equation that ignores compressive heating (i.e., the first right-hand side term in (2.9)). Equations (2.38) are a mathematically well-posed problem in fluid dynamics with any meaningful equation of state, $\rho(T, S, p)$. If compressibility is included in the equation of state, it is usually sufficiently accurate to replace p by its hydrostatic estimate, $-\rho_0 gz$ (where $-z$ is the depth beneath a mean sea level at $z = 0$), because $\delta\rho/\rho \ll 1$ for the ocean. (Equations (2.38) should not be confused with the use of the same name for the approximate equation of state (2.33). It is regrettable that history has left us with this non-unique nomenclature.)

The evolutionary equations for entropy and, using (2.33), density are redundant with (2.38):

$$T\frac{D\eta}{Dt} = \mathcal{Q} - \mu\mathcal{S}; \tag{2.39}$$

$$\frac{1}{\rho_0}\frac{D\rho}{Dt} = -\frac{\alpha}{c_p}\mathcal{Q} + \beta\mathcal{S}. \tag{2.40}$$

This type of redundancy is due to the simplifying thermodynamic approximations made here. Therefore (2.40) does not need to be included explicitly in solving (2.38) for **u**, T, and S.

Qualitatively the most important dynamical consequence of making the Boussinesq dynamical approximation in (2.38) is the exclusion of sound waves, including shock waves (cf., Section 2.3.1). Typically sound waves have relatively little energy in the ocean and atmosphere (barring asteroid impacts, volcanic eruptions, jet airplane wakes, and nuclear explosions). Furthermore, they have little influence on the evolution of larger scale, more energetic motions that usually are of more interest. The basis for the approximation that neglects sound wave dynamics, can alternatively be expressed as

$$M = \frac{V}{C_s} \ll 1. \tag{2.41}$$

C_s is the sound speed ≈ 1500 m s^{-1} in the ocean; V is a fluid velocity typically ≤ 1m s^{-1} in the ocean; and M is the *Mach number*. So $M \approx 10^{-3}$ under these conditions. In contrast, in and around stars and near jet airplanes, M is often of order one or larger.

Motions with $\mathcal{Q} = \mathcal{S} = 0$ are referred to as *adiabatic*, and motions for which this is not true are *diabatic*. The last two equations in (2.38) show that T and S are conservative tracers under adiabatic conditions; they are invariant following a material parcel when compression, mixing, and heat and water sources are negligible. Equations (2.40) and (2.41) show that η and ρ are also conservative tracers under these conditions. Another name for adiabatic motion is *isentropic* motion because the entropy does not change in the absence of sources or sinks of heat and tracers. Also, under these conditions, isentropic is the same as *isopycnal*, with the implication that parcels can move laterally on stably stratified isopycnal surfaces but not across them. The adiabatic idealization is not exactly true for the ocean, even in the stratified interior away from boundary layers (Chapter 6), but it often is nearly true over time intervals of months or even years.

2.2.3 Boundary conditions

The boundary conditions for the ocean are comprised of kinematic, continuity, and flux types (Section 2.1.3). The usual choices for oceanic models are the following (Fig. 2.5).

Sides/bottom

At $z = -H(x, y)$, there is no flow into the solid boundary, $\mathbf{u} \cdot \hat{\mathbf{n}} = 0$, which is the kinematic condition (2.14).

Sides/bottom and top

At all boundaries there is a specified tracer flux, commonly assumed to be zero at the solid surfaces (or at least negligibly small on fluid time scales that are much shorter than, say, geological time scales), but the tracer fluxes are typically non-zero at the air–sea interface. For example, although there is a geothermal flux into the ocean from the cooling of Earth's interior, it is much smaller on average (about 0.09 W (i.e., watt) m^{-2}) than the surface heat exchange with the atmosphere, typically many tens of $W\,m^{-2}$. However, in a few locations over hydrothermal vents, the geothermal flux is large enough to force upward convective plumes in the abyssal ocean.

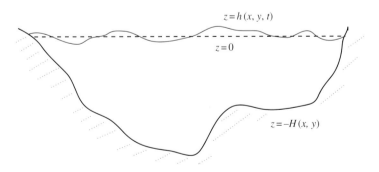

Fig. 2.5. Configuration for an oceanic domain. The heights, 0, h, and $-H$, represent the mean sea level, instantaneous local sea level, and bottom position, respectively.

At all boundaries there is a specified momentum flux: a drag stress due to currents flowing over the underlying solid surface or the wind acting on the upper free surface or relative motion between sea ice on a frozen surface and the adjacent currents. If the stress is zero the boundary condition is called *free slip*, and if the tangential relative motion is zero the condition is called *no slip*. A no-slip condition causes non-zero tangential boundary stress as an effect of viscosity acting on an adjacent fluid moving relative to the boundary.

Top

At the top of the ocean, $z = h(x, y, t)$, the kinematic free-surface condition from (2.17) is

$$w = \frac{Dh}{Dt},$$

with h the height of the ocean surface relative to its mean level of $z = 0$. The mean sea level is a hypothetical surface associated with a motionless ocean; it corresponds to a surface of constant gravitational potential – almost a sphere for Earth, even closer to an oblate spheroid with an Equatorial bulge, and actually quite convoluted due to inhomogeneities in the solid Earth with local-scale wrinkles of $\mathcal{O}(10)$ m elevation. Of course, determining h is necessarily part of an oceanic model solution.

Also at $z = h(x, y, t)$, the continuity of pressure implies that

$$p = p_{\text{atm}}(x, y, t) \approx p_{\text{atm},0}, \tag{2.42}$$

where the latter quantity is a constant $\approx 10^5 \, \text{kg m}^{-1} \text{s}^{-2}$ (or 10^5 Pa). Since $\delta p_{\text{atm}} / p_{\text{atm}} \approx 10^{-2}$, then, with a hydrostatic estimate of the oceanic pressure fluctuation at $z = 0$ (viz., $\delta p_{\text{oce}} = g \rho_0 h$), then $\delta p_{\text{oce}} / p_{\text{atm}} \approx g \rho_0 h / p_{\text{atm},0} = 10^{-2}$ for an h of only 10 cm. The latter magnitude for h is small compared to high-frequency, surface gravity wave height variations (i.e., with typical wave amplitudes of $\mathcal{O}(1)$ m and periods of $\mathcal{O}(10)$ s), but it is not necessarily small compared to the wave-averaged sea level changes associated with oceanic currents at lower frequencies of minutes and longer. However, if the surface height changes to cancel the atmospheric pressure change, with $h \approx -\delta p_{\text{atm}} / g \rho_0$ (e.g., a surface depression under high surface air pressure), the combined weight of air and water, $p_{\text{atm}} + p_{\text{oce}}$, along a horizontal surface (i.e., at constant z) is spatially and temporally uniform in the water, so no oceanic accelerations arise due to a horizontal pressure gradient force. This type of oceanic response is called the *inverse barometer response*, and it is common for slowly evolving, large-scale atmospheric pressure changes such as those in synoptic weather patterns. In nature h does vary due to surface waves, wind-forced flows, and other currents.

Rigid-lid approximation

A commonly used – and mathematically easier to analyze – alternative for the free surface conditions at the top of the ocean (the two preceding equations) is the *rigid-lid approximation* in which the boundary at $z = h$ is replaced by one at the mean sea level, $z = 0$. The approximate kinematic condition there becomes

$$w(x, y, 0, t) = 0. \tag{2.43}$$

The tracer and momentum flux boundary conditions are applied at $z = 0$. Variations in p_{atm} are neglected (mainly because they cause an inverse barometer response without causing currents except temporarily during an adjustment to the static balance), and h is no longer a *prognostic variable* of the ocean model (i.e., one where the time derivative must be integrated explicitly as an essential part of the governing partial differential equation system). However, as part of this rigid-lid approximation, a hydrostatic, *diagnostic* (i.e., referring to a dependent variable that can be evaluated in terms of the prognostic variables outside the system integration process) estimate can be made from the ocean surface pressure at the rigid lid for the implied sea-level fluctuation, h_*, and its associated vertical velocity, w_*, viz,

$$h_* \approx \frac{1}{g\rho_0} \left(p(x, y, 0, t) - p_{atm} \right), \quad w_* = \frac{Dh_*}{Dt}. \tag{2.44}$$

This approximation excludes surface gravity waves from the approximate model, but is generally quite accurate for calculating motions on larger space and slower time scales. The basis of this approximation is the relative smallness of surface height changes for the ocean, $h/H = \mathcal{O}(10^{-3}) \ll 1$, and the weakness of dynamical interactions between surface gravity waves and the larger scale, slower currents. More precisely stated, the rigid-lid approximation is derived by a Taylor series expansion of the free surface conditions around $z = 0$; e.g., the kinematic condition,

$$\frac{Dh}{Dt} = w(h) \approx w(0) + h\frac{\partial w}{\partial z}(0) + \cdots, \tag{2.45}$$

neglecting terms that are small in h_*/H, w_*/W, and h_*/WT (H, T, and W are typical values for the vertical length scale, time scale, and vertical velocity of currents in the interior). A more explicit analysis to justify the rigid-lid approximation is given near the end of Section 2.4.2 where specific estimates for T and W are available.

An ancillary consequence of the rigid-lid approximation is that mass is no longer explicitly exchanged across the sea surface since an incompressible ocean with a rigid lid has a constant volume. Instead this mass flux is represented as an exchange of chemical composition; e.g., the actual injection of fresh water that occurs by precipitation is represented as a virtual outward flux of S based upon its local diluting effect on seawater, using the relation

$$\frac{\delta H_2 O}{H_2 O} = -\frac{\delta S}{S}. \tag{2.46}$$

The denominators are the average amounts of water and salinity in the affected volume.

2.3 Atmospheric approximations

2.3.1 Equation of state for an ideal gas

Assume as a first approximation that air is an *ideal gas* with constant proportions among its primary constituents and without any water vapor, i.e., a *dry atmosphere*. (In this book we will not explicitly treat the often dynamically important effects of water in the atmosphere, thereby avoiding the whole subject of cloud effects.) Thus, p and T are the state variables, and the equation of state is

$$\rho = \frac{p}{RT}, \tag{2.47}$$

with $R = 287 \, \mathrm{m^2 \, s^{-2} \, K^{-1}}$ for the standard composition of air. The associated internal energy is $e = c_v T$, with a heat capacity at constant volume, $c_v = 717 \, \mathrm{m^2 \, s^{-2} \, K^{-1}}$. The internal energy equation (2.10) becomes

$$c_v \frac{DT}{Dt} = -p \frac{D}{Dt}\left(\frac{1}{\rho}\right) + \mathcal{Q}. \tag{2.48}$$

In the absence of other state variables influencing the entropy, (2.11) becomes

$$T\frac{D\eta}{Dt} = \mathcal{Q}, \tag{2.49}$$

and in combination with (2.48) it becomes

$$T\frac{D\eta}{Dt} = \frac{De}{Dt} + p\frac{D}{Dt}\left(\frac{1}{\rho}\right)$$

$$= c_v \frac{DT}{Dt} + p\frac{D}{Dt}\left(\frac{RT}{p}\right)$$

$$= c_p \frac{DT}{Dt} - \frac{1}{\rho}\frac{Dp}{Dt}. \tag{2.50}$$

Here $c_p = c_v + R = 1004 \, \mathrm{m^2 \, s^{-2} \, K^{-1}}$.

An alternative state variable is the *potential temperature*, θ, related to the *potential density*, ρ_{pot}, with both defined as follows:

$$\theta = T \left(\frac{p_0}{p} \right)^{\kappa}, \quad \rho_{pot} = \frac{p_0}{R\theta} = \rho \left(\frac{p_0}{p} \right)^{1/\gamma}, \tag{2.51}$$

where $\kappa = R/c_p \approx 2/7$, $\gamma = c_p/c_v \approx 7/5$, and p_0 is a reference constant for pressure at sea level, $p_{atm,0} \approx 10^5 \, \mathrm{kg\,m^{-1}\,s^{-2}} = 1 \, \mathrm{Pa}$. From (2.47)–(2.51), the following are readily derived:

$$\frac{D\theta}{Dt} = \left(\frac{p_0}{p} \right)^{\kappa} \frac{\mathcal{Q}}{c_p} = \frac{\tilde{\mathcal{Q}}}{c_p}, \tag{2.52}$$

and

$$\frac{D\rho_{pot}}{Dt} = -\frac{\rho_{pot}\mathcal{Q}}{c_p T}. \tag{2.53}$$

Thus, in isentropic (adiabatic) motions with $\mathcal{Q} = \tilde{\mathcal{Q}} = 0$, both θ and ρ_{pot} evolve as conservative tracers, but T and ρ change along trajectories due to the compression or expansion of a parcel with the pressure changes encountered *en route*. Being able to distinguish between conservative and non-conservative effects is the reason for the alternative thermodynamic variables defined in (2.51). One can similarly define θ and ρ_θ for the ocean using its equation of state; the numerical values for oceanic θ do not differ greatly from its T values, even though ρ changes much more with depth than ρ_θ because seawater density has a much greater sensitivity to compression than seawater temperature has (Fig. 2.7 below).

Sound waves

As a somewhat tangential topic, consider the propagation of *sound waves* (or *acoustic waves*) in air. With an adiabatic assumption (i.e., $\mathcal{Q} = 0$), the relation for conservation of ρ_{pot} (2.53) implies

$$\frac{D}{Dt} \left[\rho \left(\frac{p_0}{p} \right)^{1/\gamma} \right] = 0$$

$$\left(\frac{p_0}{p} \right)^{1/\gamma} \left(\frac{D\rho}{Dt} - \frac{\rho}{\gamma p} \frac{Dp}{Dt} \right) = 0$$

$$\frac{D\rho}{Dt} - C_s^{-2} \frac{Dp}{Dt} = 0, \tag{2.54}$$

with $C_s = \sqrt{\gamma p/\rho} = \sqrt{\gamma R T}$, the speed of sound in air ($\approx 300 \, \mathrm{m\,s^{-1}}$ for $T = 300$ K). Now linearize this equation plus those for continuity (2.6) and momentum (i.e., (2.2) setting $\mathbf{F} = 0$) around a reference state of $\rho = \rho_0$ and $\mathbf{u} = 0$, neglecting all terms that are quadratic in fluctuations around the static reference state:

$$\frac{\partial \rho}{\partial t} - C_s^{-2} \frac{\partial p}{\partial t} = 0$$

$$\frac{\partial \rho}{\partial t} + \rho_0 \mathbf{\nabla} \cdot \mathbf{u} = 0$$

$$\frac{\partial \mathbf{u}}{\partial t} = -\frac{1}{\rho_0} \mathbf{\nabla} p \Rightarrow \mathbf{\nabla} \cdot \frac{\partial \mathbf{u}}{\partial t} = -\frac{1}{\rho_0} \nabla^2 p. \tag{2.55}$$

The combination of these equations, ∂_t (second equation) $- \partial_t$ (first) $- \rho_0 \times$ (third), implies that

$$\frac{\partial^2 p}{\partial t^2} - C_s^2 \nabla^2 p = 0. \tag{2.56}$$

This equation has the functional form of the canonical *wave equation* that is representative of the general class of hyperbolic partial differential equations. It has general solutions of the form,

$$p(\mathbf{x}, t) = \mathcal{F}[\hat{\mathbf{e}} \cdot \mathbf{x} - C_s t], \tag{2.57}$$

when C_s is constant (i.e., assuming $T \approx T_0$). This form represents the uniform propagation of a disturbance (i.e., a weak perturbation about the reference state) having any shape (or wave form) $\mathcal{F}$ with speed C_s and an arbitrary propagation direction $\hat{\mathbf{e}}$. Equation (2.57) implies that the wave shape is unchanged with propagation. This is why sound is a reliable means of communication. Equivalently, one can say that sound waves are *non-dispersive* (Sections 3.1.2, 4.2, ff). Analogous relations can be derived for oceanic sound propagation without making the incompressibility approximation, albeit with a different thermodynamic prescription for C_s that has a quite different magnitude, $\approx 1500 \, \mathrm{m \, s^{-1}}$.

2.3.2 A stratified resting state

A *resting atmosphere*, in which $\mathbf{u} = 0$ and all other fields are horizontally uniform, $\mathbf{\nabla}_h = 0$, is a consistent solution of the conservative governing equations. The momentum equation (2.2) with $\mathbf{F} = 0$ is non-trivial only in the vertical direction, viz,

$$\frac{\partial p}{\partial z} = -g\rho. \tag{2.58}$$

This is a differential expression of hydrostatic balance. It implies that the pressure at a point is approximately equal to the vertically integrated density (i.e., the weight) for all the fluid above it, assuming that outer space is weightless. Hydrostatic balance plus the equation of state (2.47) plus the vertical profile of any thermodynamic quantity (i.e., T, p, ρ, θ, or ρ_{pot}) determines the vertical profiles of all such quantities in a resting atmosphere. (Again, there is an analogous oceanic resting state.)

One simple example is a resting *isentropic atmosphere*, in which $\theta(z, t) = \theta_0$, a constant:

$$\Rightarrow \frac{d\theta}{dz} = 0$$

$$\Rightarrow \frac{d}{dz}\left[T\left(\frac{p_0}{p}\right)^{\kappa}\right] = 0$$

$$\Rightarrow \frac{dT}{dz} = -\frac{g}{c_p} \approx -10^{-2}\,\mathrm{K\,m^{-1}}, \tag{2.59}$$

after using (2.47) and (2.58). This final relation defines the *lapse rate* of an isentropic atmosphere, also called the adiabatic lapse rate. Integrating (2.59) gives

$$T = \theta_0 - \frac{gz}{c_p} \tag{2.60}$$

if $T = \theta_0$ at $z = 0$. Thus, the air is colder with altitude as a consequence of the decreases in pressure and density. Also,

$$p = p_0 \left(\frac{T}{\theta_0}\right)^{1/\kappa}$$

$$\Rightarrow p = p_0\left[1 - \frac{gz}{c_p\theta_0}\right]^{1/\kappa} \tag{2.61}$$

and

$$\rho = \frac{p_0}{R\theta_0}\left(\frac{p}{p_0}\right)^{1/\gamma} \tag{2.62}$$

$$\Rightarrow \rho = \rho_{pot,0}\left[1 - \frac{gz}{c_p\theta_0}\right]^{1/\kappa\gamma}. \tag{2.63}$$

An isentropic atmosphere ends (i.e., $\rho = p = T = 0$) at a finite height above the ground,

$$H = \frac{c_p\theta_0}{g} \approx 3 \times 10^4\,\mathrm{m} \tag{2.64}$$

for $\theta_0 = 300\,\mathrm{K}$.

A different example of a resting atmosphere is an *isothermal atmosphere*, with $T = T_0$. From (2.47) and (2.58),

$$\frac{d\rho}{dz} = \frac{1}{RT_0}\frac{dp}{dz} = -\frac{g}{RT_0}\rho.$$

$$\Rightarrow \rho = \rho_0 e^{-z/H_0}. \tag{2.65}$$

The *scale height* for exponential decay of the density is $H_0 = RT_0/g \approx 10^4\,\text{m}$ for $T_0 = 300\,\text{K}$. Also,

$$p = RT_0\rho_0 e^{-z/H_0}, \quad \theta = T_0 e^{\kappa z/H_0}, \quad \rho_{\text{pot}} = \rho_0 e^{-\kappa z/H_0}. \tag{2.66}$$

Thus, an isothermal atmosphere extends to $z = \infty$ (ignoring any astronomical influences), and it has an increasing θ with altitude and a ρ_{pot} that decreases much more slowly than ρ (since $\kappa \ll 1$).

Earth's atmosphere has vertical profiles much closer to isothermal than isentropic in the particular sense that it is stably stratified, with $\partial_z\theta > 0$ on average. Similarly, the ocean is stably stratified on average. Figures 2.6 and 2.7 show horizontal- and time-averaged vertical profiles from measurements that can usefully be viewed as the stratified resting states around which the wind- and current-induced thermodynamic and pressure fluctuations occur.

2.3.3 Buoyancy oscillations and convection

Next, consider the adiabatic dynamics of an air parcel slightly displaced from its resting height. Denote the resting, hydrostatic profiles of pressure and density by $\overline{p}(z)$ and $\overline{\rho}(z)$ and the vertical displacement of a parcel originally at z_0 by δz. (The overbar denotes an average quantity.) The conservative vertical momentum balance (2.2) is

$$\frac{Dw}{Dt} = \frac{D^2\,\delta z}{Dt^2} = -g - \frac{1}{\rho}\frac{\partial p}{\partial z}. \tag{2.67}$$

Now make what may seem at first to be an *ad hoc* assumption: as the parcel moves, the parcel pressure, p, instantaneously adjusts to the local value of $\overline{p}$. (This assumption excludes any sound wave behavior in the calculated response; in fact, it becomes valid as a result of sound waves having been emitted in conjunction with the parcel displacement, allowing the parcel pressure to locally equilibrate.) After using the hydrostatic balance of the mean profile to substitute for $\partial_z\overline{p}(z_0 + \delta z)$,

$$\frac{D^2\,\delta z}{Dt^2} = g\left(\frac{\overline{\rho} - \rho}{\rho}\right)\Bigg|_{z=z_0+\delta z}$$

$$= g\left(\frac{\overline{\rho}_{\text{pot}} - \rho_{\text{pot}}}{\rho_{\text{pot}}}\right)\Bigg|_{z=z_0+\delta z}. \tag{2.68}$$

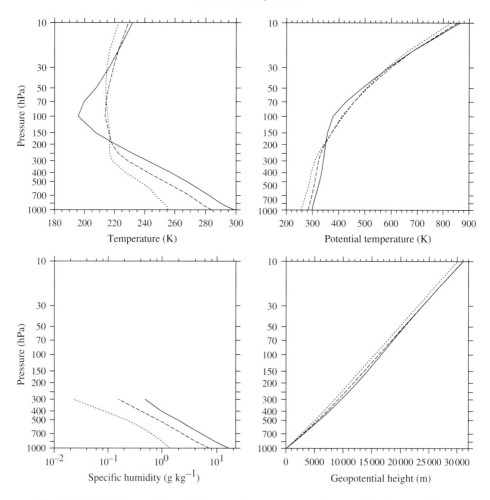

Fig. 2.6. Vertical profiles of time- and area-averaged atmospheric quantities: (upper left) temperature, T [K]; (upper right) potential temperature, θ [K]; (lower left) specific humidity, q [mass fraction $\times 10^3$]; and (lower right) geopotential height, Z [m]. The vertical axis is pressure, p [hPa $= 10^2$ Pa]. In each panel there are curves for three different areas: (solid) tropics, with latitudes $\pm(0\text{–}15)°$; (dash) middle latitudes, $\pm(30\text{–}60)°$; and (dot) poles, $\pm(75\text{–}90)°$. Note the poleward decreases in T and q; the reversal in $T(p)$ at the tropopause, $p \approx 100\text{–}200\,\text{hPa}$; the ubiquitously positive stratification in $\theta(p)$ that increases in the stratosphere; the strong decay of q with height (until reaching the stratosphere, not plotted, where it becomes more nearly uniform); and the robust, monotonic relation between Z and p. (National Centers for Environmental Prediction climatological analysis (Kalnay *et al.*, 1996), courtesy of Dennis Shea, National Center for Atmospheric Research.)

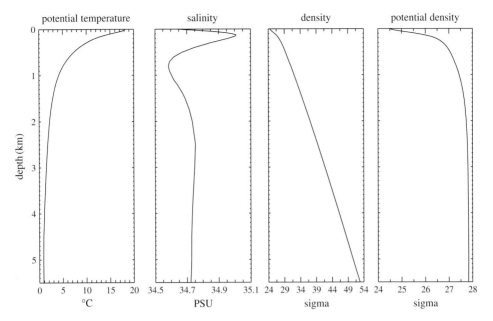

Fig. 2.7. Mean vertical profiles of θ (which is nearly the same as *in situ* T on the scale of this plot), S, ρ, and ρ_θ (i.e., potential density with a reference pressure at the surface) for the ocean. These are averages over time and horizontal position for a historical collection of hydrographic measurements (Steele *et al.*, 2001). The "sigma" unit for density is $\mathrm{kg\,m}^{-3}$ after subtracting a constant value of 10^3. Note the strongly stratified thermocline in T and pycnocline in ρ_θ; the layered influences in S of tropical precipitation excess near the surface, subtropical evaporation excess near 200 m depth, and subpolar precipitation excess near 800 m depth; the increase in ρ with depth due to compressibility, absent in ρ_θ; and the weakly stratified abyss in T, S, and ρ_θ. (Courtesy of Gokhan Danabasoglu, National Center for Atmospheric Research.)

The second line follows from ρ and $\overline{\rho}$ depending only on a common pressure-dependent factor in relation to ρ_{pot} and $\overline{\rho}_{\mathrm{pot}}$, which then cancels out between the numerator and denominator. Since potential density is preserved following a parcel for adiabatic motions and δz is small so that the potential density profile can be Taylor-expanded about the resting location of the parcel, then

$$\rho_{\mathrm{pot}}(z_0 + \delta z) = \overline{\rho}_{\mathrm{pot}}(z_0) = \overline{\rho}_{\mathrm{pot}}(z_0 + \delta z) - \delta z \frac{d\overline{\rho}_{\mathrm{pot}}}{dz}(z_0) + \cdots$$

$$\Rightarrow \frac{D^2 \delta z}{Dt^2} = -N^2(z_0)\,\delta z + \cdots, \tag{2.69}$$

where

$$N^2 = -g\frac{d}{dz}\ln[\overline{\rho}_{\mathrm{pot}}] \tag{2.70}$$

is the square of the *buoyancy frequency* or *Brünt–Väisällä frequency*. The solution
of (2.69) shows that the parcel displacement evolves in either of two ways. If $N^2 >$
0 (i.e., $\overline{\rho}_{\mathrm{pot}}$ decreases with altitude, indicative of lighter air above denser air), the
atmosphere is *stably stratified*, and the solutions are $\delta z \propto e^{\pm iNt}$ where $i = \sqrt{-1}$.
These are oscillations in the vertical position of the parcel with a period, $P =$
$2\pi/N$. The oscillations are a simple form of *internal gravity waves*. However, if
$N^2 < 0$ (with denser air above lighter air), there is a solution, $\delta z \propto e^{|N|t}$, that grows
without limit (up to a violation of the assumption of small δz) and indicates that the
atmosphere is *unstably stratified* with respect to a parcel displacement. The *growth
rate* for the instability is $|N|$, with a growth time of $|N|^{-1}$. The fluid motion that
arises from unstable stratification is called *convection* or gravitational instability.

   Using the previous relations and taking the overbar symbol as implicit,

$$
\begin{aligned}
N^2 &= -\frac{g}{\rho_{\mathrm{pot}}}\frac{\partial \rho_{\mathrm{pot}}}{\partial z} \\[2mm]
&= \frac{g}{\theta}\frac{\partial \theta}{\partial z} \\[2mm]
&= \frac{g}{T}\left(\frac{\partial T}{\partial z} - \frac{\kappa T}{p}\frac{\partial p}{\partial z}\right) \\[2mm]
&= \frac{g}{T}\left(\frac{\partial T}{\partial z} + \frac{g}{c_{\mathrm{p}}}\right) \\[2mm]
&= \frac{g}{T}\left(\frac{\partial T}{\partial z} - \left.\frac{\partial T}{\partial z}\right|_{\delta\eta=0}\right).
\end{aligned}
\tag{2.71}
$$

N is related to the difference between the actual lapse rate and the adiabatic or
isentropic rate that appears in (2.59). In the extra-tropical troposphere, a typical
value for N is about $10^{-2}\,\mathrm{s}^{-1} \Rightarrow P \approx 10$ min; in the stratosphere, N is larger (NB,
the increase in $d_z\theta$ above the tropopause; Fig. 2.6). There are analogous relations
for the ocean based on its equation of state. A typical upper-ocean value for N
is similar in magnitude, $\sim 10^{-2}\,\mathrm{s}^{-1}$, within the pycnocline underneath the often
well-mixed surface boundary layer where T and S are nearly uniform and $N \approx 0$;
in the abyssal ocean N^2 values are usually positive but much smaller than in the
upper oceanic pycnocline (Fig. 2.7).

2.3.4 Hydrostatic balance

The hydrostatic relation (2.58) is an exact one for a resting atmosphere. But it
is also approximately valid for fluid motions superimposed on mean profiles of
$\overline{p}(z)$ and $\overline{\rho}(z)$ if the motions are "thin" (i.e., have a small aspect ratio, $H/L \ll 1$,

where H and L are typical vertical and horizontal length scales). All large-scale motions are thin, insofar as their L is larger than the depth of the ocean ($\approx 5\,\text{km}$) or height of the troposphere ($\approx 10\,\text{km}$). This is demonstrated with a *scale analysis* of the vertical component of the momentum equation (2.2). If V is a typical horizontal velocity, then $W \sim VH/L$ is a typical vertical velocity such that the contributions to δ are similar for all coordinate directions. Assume that the advective acceleration and pressure gradient terms have comparable magnitudes in the horizontal momentum equation, i.e.,

$$\frac{D\mathbf{u}_\text{h}}{Dt} \sim \frac{1}{\rho}\mathbf{V}_\text{h}p$$

(NB, the subscript h denotes horizontal component). For $\rho \approx \overline{\rho} \sim \rho_0$ and $t \sim L/V$ (advective scaling; Section 2.1.1), this implies that the pressure fluctuations have a scaling estimate of $\delta p \sim \rho_0 V^2$. The further assumption that density fluctuations have a size consistent with these pressure fluctuations through hydrostatic balance implies that $\delta \rho \sim \rho_0 V^2/gH$. The hydrostatic approximation to (2.2) requires that

$$\rho \frac{Dw}{Dt} \ll \delta p_z \sim g\,\delta\rho$$

in the vertical momentum balance. Using the preceding scale estimates, the left- and right-hand sides of this inequality are estimated as

$$\rho_0 \cdot \frac{V}{L} \cdot \frac{VH}{L} \ll \rho_0 \frac{V^2}{H},$$

or, dividing by the right-hand side quantities,

$$\left(\frac{H}{L}\right)^2 \ll 1. \tag{2.72}$$

This is the condition for validity of the hydrostatic approximation for a non-rotating flow (cf., (2.111)), and it must be satisfied for large-scale flows because of their thinness.

2.3.5 Pressure coordinates

With the hydrostatic approximation (2.58), almost all aspects of the fully compressible atmospheric dynamics can be made implicit by transforming the equations to *pressure coordinates*. Formally this transformation from "height" or "physical" coordinates $(\mathbf{x}, t)$ to pressure coordinates $(\tilde{\mathbf{x}}, \tilde{t})$ is defined by

$$\tilde{x} = x, \quad \tilde{y} = y, \quad \tilde{z} = F(p), \quad \tilde{t} = t; \tag{2.73}$$

F can be any monotonic function. In height coordinates z is an independent variable, while $p(x, y, z, t)$ is a dependent variable; in pressure coordinates, $\tilde{z}(p)$ is independent, while $z(\tilde{x}, \tilde{y}, \tilde{z}, \tilde{t})$ is dependent. The pressure–height relationship is a monotonic one (Fig. 2.6, lower right) because of nearly hydrostatic balance in the atmosphere. Monotonicity is a necessary condition for $F(p)$ to be a valid alternative coordinate.

Meteorological practice includes several alternative definitions of F; two common ones are

$$F(p) = \frac{(p_0 - p)}{g \rho_0}, \tag{2.74}$$

and

$$F(p) = H_0 \left(1 - \left[\frac{p}{p_0} \right]^\kappa \right), \qquad H_0 = \frac{c_p T_0}{g} \; (\approx 30 \,\text{km}). \tag{2.75}$$

Both of these functions have units of height [m]. They have the effect of transforming a possibly infinite domain in z into a finite one in $\tilde{z}$, whose outer boundary condition is $p \to 0$ as $z \to \infty$. The second choice yields $\tilde{z} = z$ for $z \leq H_0$ in the special case of an isentropic atmosphere (2.61). The resulting equations are similar in their properties with either choice of F, but (2.75) is the one used in (2.76) ff.

The transformation rules for derivatives when only the $\tilde{z}$ coordinate is redefined (as in (2.73)) are the following:

$$\partial_x = \partial_{\tilde{x}} + \frac{\partial \tilde{z}}{\partial x} \partial_{\tilde{z}} = \partial_{\tilde{x}} - \frac{\partial_{\tilde{x}} z}{\partial_{\tilde{z}} z} \partial_{\tilde{z}}$$

$$\partial_y = \partial_{\tilde{y}} + \frac{\partial \tilde{z}}{\partial y} \partial_{\tilde{z}} = \partial_{\tilde{y}} - \frac{\partial_{\tilde{y}} z}{\partial_{\tilde{z}} z} \partial_{\tilde{z}}$$

$$\partial_z = \frac{\partial \tilde{z}}{\partial z} \partial_{\tilde{z}} = \frac{1}{\partial_{\tilde{z}} z} \partial_{\tilde{z}}$$

$$\partial_t = \partial_{\tilde{t}} + \frac{\partial \tilde{z}}{\partial t} \partial_{\tilde{z}} = \partial_{\tilde{t}} - \frac{\partial_{\tilde{t}} z}{\partial_{\tilde{z}} z} \partial_{\tilde{z}}. \tag{2.76}$$

The relations between the first and center column expressions in (2.76) are the result of applying the chain rule of calculus; e.g., the first line results from applying $\partial_x |_{y,z,t}$ to a function whose arguments are $(\tilde{x}(x), \tilde{y}(y), \tilde{z}(x, y, z, t), \tilde{t}(t))$. The coefficient factors in the third column of the equations in (2.76) are derived by applying the first two columns to the quantity z; e.g.,

$$\frac{\partial z}{\partial x} = \frac{\partial z}{\partial \tilde{x}} + \frac{\partial \tilde{z}}{\partial x} \frac{\partial z}{\partial \tilde{z}} = 0$$

$$\Rightarrow \frac{\partial \tilde{z}}{\partial x} = -\frac{\partial_{\tilde{x}} z}{\partial_{\tilde{z}} z}. \tag{2.77}$$

The substantial derivative has the same physical meaning in either coordinate system because the rate of change with time following the flow is independent of the spatial coordinate system it is evaluated in. It also has a similar mathematical structure in any space-time coordinate system:

$$\frac{D}{Dt} = \frac{Dt}{Dt}\partial_t + \frac{Dx}{Dt}\partial_x + \frac{Dy}{Dt}\partial_y + \frac{Dz}{Dt}\partial_z$$

$$= \partial_t + u\partial_x + v\partial_y + w\partial_z$$

$$= \frac{D\tilde{t}}{Dt}\partial_{\tilde{t}} + \frac{D\tilde{x}}{Dt}\partial_{\tilde{x}} + \frac{D\tilde{y}}{Dt}\partial_{\tilde{y}} + \frac{D\tilde{z}}{Dt}\partial_{\tilde{z}}$$

$$= \partial_{\tilde{t}} + u\partial_{\tilde{x}} + v\partial_{\tilde{y}} + \omega\partial_{\tilde{z}}. \qquad (2.78)$$

The first two lines are expressed as applicable to a function in height coordinates and the last two to a function in pressure coordinates. What is $\omega = D_t\tilde{z}$? By using the right-hand side of the transformation rules (2.76) substituted into this expression and the second line in (2.78), the expression for ω is derived to be

$$\omega = \frac{1}{\partial_z z}\left(w - \frac{\partial z}{\partial \tilde{t}} - u\frac{\partial z}{\partial \tilde{x}} - v\frac{\partial z}{\partial \tilde{y}}\right). \qquad (2.79)$$

The physical interpretation of ω is the rate of fluid motion across a surface of constant pressure (i.e., an *isobaric surface*, $\tilde{z} = $ const.), which itself is moving in physical space. Stated more literally, it is the rate at which the coordinate $\tilde{z}$ changes following the flow.

Now consider the equations of motion in the transformed coordinate frame. The hydrostatic relation (2.58), with (2.47), (2.51), (2.75), and (2.76), becomes

$$\frac{\partial \Phi}{\partial \tilde{z}} = \frac{c_p}{H_0}\theta, \qquad (2.80)$$

where

$$\Phi = gz \qquad (2.81)$$

is the geopotential function appropriate to the pressure-coordinate frame. The substantial time derivative is interpreted as the final line of (2.78). After similar manipulations, the horizontal momentum equation from (2.2) becomes

$$\frac{D\mathbf{u}_h}{Dt} = -\tilde{\mathbf{V}}_h\Phi + \mathbf{F}_h; \qquad (2.82)$$

the subscript h again denotes horizontal component. The internal energy equation is the same as (2.52), namely,

$$\frac{D\theta}{Dt} = \frac{\tilde{Q}}{c_p}, \qquad (2.83)$$

with

$$\tilde{Q} = \left(\frac{p_0}{p}\right)^{\kappa} \quad Q = \frac{\varOmega}{1 - \tilde{z}/H_0},$$ (2.84)

the potential temperature heating rate. The continuity equation (2.6) becomes

$$\tilde{\nabla}_{h} \cdot \mathbf{u}_{h} + \frac{1}{G(\tilde{z})} \frac{\partial}{\partial \tilde{z}} \left[G(\tilde{z})\omega\right] = 0,$$ (2.85)

with the variable coefficient,

$$G(\tilde{z}) = \left(1 - \frac{\tilde{z}}{H_0}\right)^{(1-\kappa)/\kappa}.$$ (2.86)

Note that (2.85) does not have any time-dependent term expressing the compressibility of a parcel. The physical reason is that the transformed coordinates have an elemental "volume" that is not a volume in physical space,

$$d\,\mathrm{vol} = dx\,dy\,dz,$$

but a mass amount,

$$dx\,dy\,dp = dx\,dy\,p_z\,dz = -g\rho\,d\,\mathrm{vol} \propto d\,\mathrm{mass},$$

when the hydrostatic approximation is made. With the assumption that mass is neither created nor destroyed, the pressure-coordinate element does not change with time.

   The equation set (2.80)–(2.85) is called the (hydrostatic) *primitive equations* (PE). (Similar to the Boussinseq equations (2.38), the primitive equations often are rewritten in a rotating coordinate frame with the additional Coriolis force, $-2\varOmega\mathbf{z} \times \mathbf{u}_{h}$, in (2.82); Section 2.4.) It comprises a closed set for the dependent variables Φ, θ, $\mathbf{u}_{h}$, and ω. It can be augmented by various diagnostic equations – such as (2.75) for p, (2.47) for ρ, (2.79) for w – when these other quantities are of interest. Its solutions can also be transformed back into height coordinates by (2.73) for a geographical interpretation in physical space. The same name and its abbreviation, PE, is used for the simplified form of the Boussinesq equations (2.38) with an additional hydrostatic approximation, and the modifiers "in physical coordinates" or "in pressure coordinates" can be appended to distinguish them.

   Notice that (2.80)–(2.85) are very close in mathematical form to incompressible fluid dynamics, most specifically because there is no time derivative in (2.85): the mass conservation equation has changed its character from a prognostic to a diagnostic relation. In fact, (2.80)–(2.85) is isomorphic to a subset of the incompressible Boussinesq equations (2.38) (i.e., with the hydrostatic approximation and disregarding S for the atmosphere) if $(\partial_{\tilde{z}}G)/G$ is neglected relative to $(\partial_{\tilde{z}}\omega)/\omega$ in

(2.85). This latter approximation is appropriate whenever the actual transformed vertical scale of the motion is small in the sense of

$$\tilde{H} \ll \left(\frac{\kappa}{1-\kappa}\right) H_0 \approx 12\,\text{km}.$$

Even for troposphere-filling motions (with a vertical extent ~ 10 km), this approximation is often made for simplicity, although in practice it does not significantly complicate solving the equations. So the hydrostatic, incompressible primitive equations are one of the most fundamental equation sets for GFD studies of large-scale (thin) oceanic and atmospheric motions, and it is justified through the arguments leading either to (2.38) with (2.58) or to (2.80)–(2.85) with $G \approx 1$ or not.

The standard oceanic and atmospheric *general circulation models* – used to calculate the weather and climate – are based upon the PE. For quantitative realism, the oceanic general circulation models do include the effects of salinity, S, and the general equation of state for seawater, and the atmospheric general circulation models include water vapor, q, and $G \neq 1$. Choosing the primitive equations excludes sound waves and inaccurately represents motions with strong vertical acceleration. Examples of the latter are strong convection, surface gravity waves, and even high-frequency internal gravity waves (with frequency near N; Section 4.2.2). Since these phenomena do occur in nature, the presumption must be that these excluded or distorted motions either do not matter for the general circulation or that their important effects in the general circulation will be expressed through parameterizations (Chapter 1).

2.4 Earth's rotation

Earth and most other astronomical bodies are rotating. It is usually easier to analyze the fluid dynamics by making a transformation into the rotating reference frame of an observer on the rotating body since the relative motions are much smaller than the absolute ones; i.e., $V \ll \Omega a$, where V is a typical scale for the relative motion, Ω is the planetary rotation rate $= 2\pi\,\text{rad}\,\text{d}^{-1} \approx 0.73 \times 10^{-4}\,\text{s}^{-1}$, and $a \approx 6400$ km is its radius. If the body has an approximately spherical shape with a surrounding, gravitationally bound fluid layer, then the transformation is most appropriately done with spherical shell coordinates. (For Earth the fluid shell is quite thin when viewed on the planetary scale.) But the essential results can more simply be demonstrated for the situation of rotation in a Cartesian coordinate frame about an axis aligned with the local vertical direction (parallel to gravity). Small-scale motions typically are not influenced very much by rotation, because their time scale is short compared to the rotation period (hence their Rossby number, *Ro*, is large; see (2.102) below). Large-scale motions are influenced by rotation, but due to their thinness (i.e., small aspect ratio, (2.72)), usually only the vertical component of Earth's rotation vector is dynamically important (as explained at the end of Section 2.4.2). Compared to the true rotation vector of Earth, $\mathbf{\Omega}_e$ – with its direction parallel to the axis of rotation, pointed outward through the north pole, and magnitude equal to the angular frequency of rotation – only the local vertical component is retained here,

$$\mathbf{\Omega} = \Omega \hat{\mathbf{z}} = (\mathbf{\Omega}_e \cdot \hat{\mathbf{z}})\hat{\mathbf{z}} = |\mathbf{\Omega}_e| \sin[\theta]\hat{\mathbf{z}}; \tag{2.87}$$

θ is the latitude (Fig. 2.8). Note that Ω has different signs in the two hemispheres; by a right-handed convention (i.e., for the right thumb aligned with Earth's rotation vector, the direction of rotation coincides with the curl of the fingers), Ω is positive in the northern hemisphere since the $\mathbf{\Omega}_e \cdot \hat{\mathbf{z}}$ is positive there.

Along with the vertical-component approximation in (2.87), a spatially local approximation is also often made using a Taylor series expansion in $\theta - \theta_0 \ll 1$, or equivalently $(y - y_0) \ll a$, where a is Earth's radius:

$$\Omega = |\mathbf{\Omega}| = |\mathbf{\Omega}_e| (\sin[\theta_0] + \cos[\theta_0](\theta - \theta_0) + \cdots)$$

$$= \frac{1}{2} (f_0 + \beta_0(y - y_0) + \cdots). \tag{2.88}$$

The *Coriolis frequency* and its gradient are defined as

$$f_0 = 2|\mathbf{\Omega}_e| \sin[\theta_0] \quad \text{and} \quad \beta_0 = \frac{2|\mathbf{\Omega}_e|}{a} \cos[\theta_0]. \tag{2.89}$$

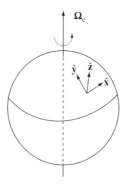

Fig. 2.8. Geometry of Earth's rotation vector, $\mathbf{\Omega}_e$. Its direction is outward and parallel to the north pole. A local Cartesian coordinate system has unit vectors, $(\hat{\mathbf{x}}, \hat{\mathbf{y}}, \hat{\mathbf{z}})$, and the local vertical component of $\mathbf{\Omega}_e$ is $\mathbf{\Omega} = |\mathbf{\Omega}_e| \sin[\theta]\hat{\mathbf{z}}$, where θ is the latitude.

f_0 changes sign between the hemispheres, vanishes at the Equator, and is largest at the poles. β_0 is positive everywhere and is largest at the Equator. When the characteristic length scale, L, is sufficiently small, then only the first term in (2.88) is retained; this is called the *f-plane* approximation. When L/a is not completely negligible, then the second term is also retained in the *β-plane* approximation. When L/a is not small, then no approximation to Ω is warranted, and a Cartesian coordinate frame is not apt.

2.4.1 Rotating coordinates

As in (2.73) for pressure coordinates, a coordinate transformation is defined here from non-rotating (or inertial) coordinates, $(\mathbf{x}, t)$, to rotating ones, $(\mathbf{X}, T)$ (Fig. 2.9). The rotating coordinates are defined by

$$X = x\cos[\Omega t] + y\sin[\Omega t],$$
$$Y = -x\sin[\Omega t] + y\cos[\Omega t],$$
$$Z = z, \quad T = t. \tag{2.90}$$

The unit vectors in the rotating frame are

$$\hat{\mathbf{X}} = \hat{\mathbf{x}}\cos[\Omega t] + \hat{\mathbf{y}}\sin[\Omega t], \quad \hat{\mathbf{Y}} = -\hat{\mathbf{x}}\sin[\Omega t] + \hat{\mathbf{y}}\cos[\Omega t],$$
$$\hat{\mathbf{Z}} = \hat{\mathbf{z}}, \tag{2.91}$$

and the associated velocity is

$$\mathbf{U} = \mathbf{u} - \Omega\hat{\mathbf{z}} \times \mathbf{x}, \tag{2.92}$$

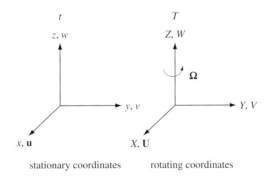

Fig. 2.9. A rotating coordinate frame with coordinates, (X, Y, Z, T), and a non-rotating frame with coordinates, (x, y, z, t). The rotation vector is parallel to the vertical axis, $\mathbf{\Omega} = \Omega\hat{\mathbf{z}}$.

where **U** is the relative velocity in the rotating reference frame. This says that the relative velocity in the rotating frame is the stationary-frame velocity minus the motion of the rotating frame itself. In terms of the velocity components,

$$U = (u + \Omega y)\cos[\Omega t] + (v - \Omega x)\sin[\Omega t],$$

$$V = -(u + \Omega y)\sin[\Omega t] + (v - \Omega x)\cos[\Omega t],$$

$$W = w. \tag{2.93}$$

Now analyze how the relevant operators are transformed. An operator in the non-rotating, or stationary, frame is denoted by a subscript s and one in the rotating frame by subscript r. The advection and gradient operators have isomorphic forms in the two frames, as can be verified by substituting from (2.90)–(2.93) and applying the differential chain rules (as in Section 2.3.5).

$$\frac{D}{Dt}_s = \partial_t + u\partial_x + v\partial_y + w\partial_z$$

$$= \frac{D}{Dt}_r = \partial_T + U\partial_X + V\partial_Y + W\partial_Z. \tag{2.94}$$

$$\mathbf{V}_s = \hat{\mathbf{x}}\partial_x + \hat{\mathbf{y}}\partial_y + \hat{\mathbf{z}}\partial_z$$

$$= \mathbf{V}_r = \hat{\mathbf{X}}\partial_X + \hat{\mathbf{Y}}\partial_Y + \hat{\mathbf{Z}}\partial_Z. \tag{2.95}$$

Similarly, the incompressible continuity equation in (2.38) preserves its form,

$$\mathbf{V}_s \cdot \mathbf{u} = \mathbf{V}_r \cdot \mathbf{U} = 0, \tag{2.96}$$

implying that material parcel volume elements are the same in each frame, with $d\mathbf{x} = d\mathbf{X}$. The tracer equations in (2.38) also preserve their form because of (2.94). The material acceleration transforms as

$$\frac{D\mathbf{u}}{Dt}_s = \frac{D}{Dt}[\hat{\mathbf{x}}u + \hat{\mathbf{y}}v + \hat{\mathbf{z}}w]$$

$$= \frac{D}{Dt}[\hat{\mathbf{X}}(U - \Omega Y) + \hat{\mathbf{Y}}(V + \Omega X) + \hat{\mathbf{Z}}W]$$

$$= \frac{D\mathbf{U}}{Dt}_r + 2\Omega\hat{\mathbf{Z}} \times \mathbf{U} + \frac{1}{\rho_0}\mathbf{V}_r P, \tag{2.97}$$

with

$$P = -\frac{\rho_0\Omega^2}{2}(X^2 + Y^2). \tag{2.98}$$

The step from the first and second lines in (2.97) is an application of (2.92). In the step to the third line, use is made of (2.94) and the relations,

$$\frac{D\hat{\mathbf{X}}}{Dt} = \Omega\hat{\mathbf{Y}}, \qquad \frac{D\hat{\mathbf{Y}}}{Dt} = -\Omega\hat{\mathbf{X}}, \qquad \frac{D\hat{\mathbf{Z}}}{Dt} = 0, \tag{2.99}$$

that describe how the orientation of the transformed coordinates rotates. Since $\mathbf{V}_s\phi = \mathbf{V}_r\phi$ by (2.95), the momentum equation in (2.38) transforms into

$$\frac{D\mathbf{U}}{Dt}_r + 2\Omega\hat{\mathbf{Z}} \times \mathbf{U} = -\mathbf{V}_r\left(\phi + \frac{P}{\rho_0}\right) - \hat{\mathbf{Z}}\frac{g\rho}{\rho_0} + \mathbf{F}. \qquad (2.100)$$

After absorbing the incremental centrifugal force potential, P/ρ_0, into a redefined geopotential function, ϕ, then (2.100) has almost the same mathematical form as the original non-rotating momentum equation, albeit in terms of its transformed variables, except for the addition of the *Coriolis force*, $-2\,\mathbf{\Omega} \times \mathbf{U}$. The Coriolis force has the effect of accelerating a rotating-frame horizontal parcel displacement in the horizontally perpendicular direction (i.e., to the right when $\Omega > 0$). This acceleration is only an apparent force from the perspective of an observer in the rotating frame, since it is absent in the inertial-frame momentum balance.

Hereafter, the original notation (e.g., $\mathbf{x}$) will also be used for rotating coordinates, and the context will make it clear which reference frame is being used. Alternative geometrical and heuristic discussions of this transformation are in Pedlosky (Chapter 1.6, 1987), Gill (Chapter 4.5, 1982), and Cushman-Roisin (Chapter 2, 1994).

2.4.2 Geostrophic balance

The *Rossby number*, *Ro*, is a non-dimensional scaling estimate for the relative strengths of the advective and Coriolis forces:

$$\frac{\mathbf{u}\cdot\mathbf{\nabla u}}{2\mathbf{\Omega}\times\mathbf{u}} \sim \frac{VV/L}{2\Omega V} = \frac{V}{2\Omega L}, \qquad (2.101)$$

or

$$Ro = \frac{V}{fL}, \qquad (2.102)$$

where $f = 2\Omega$ is the Coriolis frequency. In the ocean mesoscale eddies and strong currents (e.g., the Gulf Stream) typically have $V \leq 0.5\,\mathrm{m\ s^{-1}}$, $L \approx 50\,\mathrm{km}$, and $f \approx 10^{-4}\,\mathrm{s^{-1}}(\sim 2\pi\,\mathrm{d^{-1}})$; thus, $Ro \leq 0.1$. In the atmosphere the Jet Stream and synoptic storms typically have $V \leq 50\,\mathrm{m\ s^{-1}}$, $L \approx 10^3\,\mathrm{km}$, and $f \approx 10^{-4}\,\mathrm{s^{-1}}$; thus, $Ro \leq 0.5$. Therefore, large-scale motions have moderate or small Ro, hence strong rotational influences on their dynamics. Motions on the planetary scale have a larger $L \sim a$ and usually a smaller V, so their Ro values are even smaller.

Assume as a starting model the rotating primitive equations with the hydrostatic approximation (2.58). If $t \sim L/V \sim 1/fRo$, $\mathbf{F} \sim RofV$ (or smaller), and $Ro \ll 1$, then the horizontal velocity is approximately equivalent to the *geostrophic veloc-ity*, $\mathbf{u}_g = (u_g, v_g, 0)$, namely,

$$\mathbf{u}_h \approx \mathbf{u}_g,$$

and the horizontal component of (2.100) becomes

$$fv_g = \frac{\partial \phi}{\partial x}, \quad fu_g = -\frac{\partial \phi}{\partial y}, \tag{2.103}$$

with errors $\mathcal{O}(Ro)$. This is called *geostrophic balance*, and it defines the geostrophic velocity in terms of the pressure gradient and Coriolis frequency. The accompanying vertical force balance is hydrostatic,

$$\frac{\partial \phi}{\partial z} = -g\frac{\rho}{\rho_0} = -g(1 - \alpha\theta), \tag{2.104}$$

expressed here as a notational hybrid of (2.33), (2.58), and (2.80) with the simple equation of state,

$$\rho/\rho_0 = 1 - \alpha\theta.$$

Combining (2.103) and (2.104) yields

$$f\frac{\partial v_g}{\partial z} = g\alpha\frac{\partial \theta}{\partial x}, \quad f\frac{\partial u_g}{\partial z} = -g\alpha\frac{\partial \theta}{\partial y}, \tag{2.105}$$

called *thermal-wind balance*. Thermal-wind balance implies that the vertical gradient of horizontal velocity (or vertical *shear*) is directed along isotherms in a horizontal plane with a magnitude proportional to the horizontal thermal gradient.

Geostrophic balance implies that the horizontal velocity, $\mathbf{u}_g$, is approximately along isolines of the geopotential function (i.e., isobars) in horizontal planes. Comparing this with the incompressible velocity potential representation (Section 2.2.1) shows that

$$\psi = \frac{1}{f}\phi + \mathcal{O}(Ro); \tag{2.106}$$

i.e., the geopotential is a horizontal streamfunction whose isolines are streamlines (Section 2.1.1). For constant f (i.e., the f-plane approximation), $\partial_x u_g + \partial_y v_g = 0$ for a geostrophic flow; hence, $\delta_h = 0$ and $w = 0$ at this order of approximation for an incompressible flow. So there is no divergent potential as part of the geostrophic velocity, i.e., $X = \chi = 0$ (Section 2.2.1). However, the dynamically consistent evolution of a geostrophic flow does induce small but non-zero X, χ, and w fields associated with an *ageostrophic velocity* component that is an $\mathcal{O}(Ro)$ correction to the geostrophic flow, but the explanation for this is deferred to the topic of quasigeostrophy in Chapter 4.

Now make a scaling analysis in which the magnitudes of various fields are estimated in terms of the typical magnitudes of a few primary quantities and assumptions about what the dynamical balances are. The way that it is done here is called *geostrophic scaling*. The primary scales are assumed to be

$$u, v \sim V, \quad x, y \sim L, \quad z \sim H, \quad f \sim f_0. \tag{2.107}$$

From these additional scaling estimates are derived,

$$T \sim \frac{L}{V}, \quad p \sim \rho_0 f_0 VL, \quad \rho \sim \frac{\rho_0 f_0 VL}{gH}, \tag{2.108}$$

by advection as the dominant rate for the time evolution, geostrophic balance, and hydrostatic balance, respectively. For the vertical velocity, the scaling estimate from 3D continuity is $W \sim VH/L$. However, since geostrophic balance has horizontal velocities that are approximately horizontally non-divergent (i.e., $\mathbf{V}_h \cdot \mathbf{u}_h = 0$), they cannot provide a balance in continuity to a w with this magnitude. Therefore, the consistent w scaling must be an order smaller in the expansion parameter, Ro, namely,

$$W \sim Ro\frac{VH}{L} = \frac{V^2 H}{f_0 L^2}. \tag{2.109}$$

Similarly, by assuming that changes in $f(y)$ are small on the horizontal scale of interest (cf., (2.88)) so that they do not contribute to the leading-order momentum balance (2.103), then

$$\beta = \frac{df}{dy} \sim Ro\frac{f_0}{L} = \frac{V}{L^2}. \tag{2.110}$$

This condition for neglecting β can be recast, using $\beta \sim f_0/a$ (with $a \approx 6.4 \times 10^6$ m, Earth's radius), as a statement that $L/a = Ro \ll 1$, i.e., L is a sub-global scale. Finally, with geostrophic scaling the condition for validity of the hydrostatic approximation in the vertical momentum equation can be shown to be

$$Ro^2 \left(\frac{H}{L}\right)^2 \ll 1 \tag{2.111}$$

by an argument analogous to the non-rotating one in Section 2.3.4 (cf., (2.72)).

Equipped with these geostrophic scaling estimates, now reconsider the basis for the oceanic rigid-lid approximation (Section 2.3.3). The approximation is based on the smallness of $D_t h$ compared to interior values of w. The scalings are based on horizontal velocity, V, horizontal length, L, vertical length, H, and Coriolis frequency, f, a geostrophic estimate for the sea level fluctuation, $h \sim fVL/g$, and the advective estimate, $D_t \sim V/L$. These combine to give $D_t h \sim fV^2/g$. The geostrophic estimate for w is (2.109). So the rigid-lid approximation is accurate if

$$w \gg \frac{Dh}{Dt}$$

$$\frac{V^2 H}{f_0 L^2} \gg \frac{fV^2}{g}$$

$$R_e^2 \gg L^2, \tag{2.112}$$

with

$$R_e = \frac{\sqrt{gH}}{f}. \tag{2.113}$$

R_e is called the external or barotropic *deformation radius* (cf., Chapter 4), and it is associated with the density jump across the oceanic free surface (as opposed to the baroclinic deformation radii associated with the interior stratification; cf., Chapter 5). For mid-ocean regions with $H \approx 5000\,\text{m}$, R_e has a magnitude of several $1000\,\text{km}$. This is much larger than the characteristic horizontal scale, L, for most oceanic currents.

Geostrophic scaling analysis can also be used to determine the conditions for consistently neglecting the horizontal component of the local rotation vector, $f_h = 2\Omega_e \cos[\theta]$, compared to the local vertical component, $f = 2\Omega_e \sin[\theta]$ (Fig. 2.8). The Coriolis force in local Cartesian coordinates on a rotating sphere is

$$2\mathbf{\Omega}_e \times \mathbf{u} = \hat{x}(f_h w - fv) + \hat{y} fu - \hat{z} f_h u. \tag{2.114}$$

In the $\hat{x}$ momentum equation, $f_h w$ is negligible compared to fv if

$$Ro \frac{H}{L} \frac{f_h}{f} \ll 1, \tag{2.115}$$

based on the geostrophic scale estimates for v and w. In the $\hat{z}$ momentum equation, $f_h u$ is negligible compared to $\partial_z p/\rho_0$ if

$$\frac{H}{L} \frac{f_h}{f} \ll 1, \tag{2.116}$$

based upon the geostrophic pressure scale, $p \sim \rho_0 f L V$. In middle and high latitudes, $f_h/f \leq 1$, but it becomes large near the Equator. So, for a geostrophic flow with $Ro \leq \mathcal{O}(1)$, with small aspect ratio, and away from the Equator, the dynamical effect of the horizontal component of the Coriolis frequency, f_h, is negligible. Recall that thinness is also the basis for consistent hydrostatic balance. For more isotropic motions (e.g., in a turbulent Ekman boundary layer; Section 6.1) or flows very near the Equator, where $f \ll f_h$ since $\theta \ll 1$, the neglection of f_h is not always valid.

2.4.3 Inertial oscillations

There is a special type of horizontally uniform solution of the rotating primitive equations (either stably stratified or with uniform density); with no pressure or density variations around the resting state, no vertical velocity, and no non-conservative effects:

$$\delta\phi = \delta\theta = w = \mathbf{F} = \mathcal{Q} = \mathbf{V}_h = 0. \tag{2.117}$$

The horizontal component of (2.100) implies

$$\frac{\partial u}{\partial t} - fv = 0, \quad \frac{\partial v}{\partial t} + fu = 0, \tag{2.118}$$

and the other dynamical equations are satisfied trivially by (2.117). A linear combination of the separate equations in (2.118) as ∂_t (first) $+ f \times$ (second) yields the composite equation,

$$\frac{\partial^2 u}{\partial t} + f^2 u = 0. \tag{2.119}$$

This has a general solution,

$$u = u_0 \cos[ft + \lambda_0]. \tag{2.120}$$

Here u_0 and λ_0 are amplitude and phase constants. From the first equation in (2.118), the associated northward velocity is

$$v = -u_0 \sin[ft + \lambda_0]. \tag{2.121}$$

The solution (2.120) and (2.121) is called an *inertial oscillation*, with a period $P = 2\pi/f \approx 1\,\mathrm{d}$, varying from half a day at the poles to infinity at the Equator. Its dynamics are somewhat similar to *Foucault's pendulum* that appears to a ground-based observer to precess with frequency f as Earth rotates underneath it; but the analogy is not exact (Cushman-Roisin, Section 2.5, 1994). Durran (1993) interprets an inertial oscillation as a trajectory with constant absolute angular momentum about the axis of rotation (cf., Section 3.3.2).

For such a solution, the streamlines are parallel, and they rotate clockwise/counter-clockwise with frequency $|f|$ for $f > 0/< 0$ in the northern/southern hemisphere when viewed from above. The associated streamfunction (Section 2.2.1) is

$$\psi(x, y, t) = -u_0 \left(x \sin[ft + \lambda_0] + y \cos[ft + \lambda_0] \right). \tag{2.122}$$

The trajectories are circles (going clockwise for $f > 0$) with a radius of u_0/f, often called *inertial circles*. This direction of rotary motion is also called *anticyclonic* motion, meaning rotation in the opposite direction from Earth's rotation (i.e., with an angular frequency about $\hat{\mathbf{z}}$ with the opposite sign of f). *Cyclonic* motion is rotation with the same sign as f. The same terminology is applied to flows with the opposite or same sign, respectively, of the vertical vorticity, ζ^z, relative to f (Chapter 3).

Since $f \sim \Omega \approx 10^{-4}\,\mathrm{s}^{-1}$, it is commonly true that $f \ll N$ in the atmospheric troposphere and stratosphere, and the oceanic pycnocline. Inertial oscillations are typically slower than buoyancy oscillations (Section 2.2.3), but both are typically faster than the advective evolutionary rate, V/L, for geostrophic winds and currents.

3

Barotropic and vortex dynamics

The ocean and atmosphere are full of *vortices*, i.e., locally recirculating flows with approximately circular streamlines and trajectories. Most often the recirculation is in horizontal planes, perpendicular to the gravitational acceleration and rotation vectors in the vertical direction. Vortices are often referred to as *coherent structures*, connoting their nearly universal circular flow pattern, no matter what their size or intensity, and their longevity in a Lagrangian coordinate frame that moves with the larger scale, ambient flow. Examples include winter cyclones, hurricanes, tornadoes, dust devils, Gulf-Stream Rings, Meddies (a sub-mesoscale, sub-surface vortex, with $L \sim 10 \mathrm{s\,km}$, in the North Atlantic whose core water has chemical properties characteristic of the Mediterranean outflow into the Atlantic), and many others without familiar names. A coincidental simultaneous occurrence of well formed vortices in the Davis Strait is shown in Fig. 3.1. The three oceanic anticyclonic vortices on the southwestern side are made visible by the pattern of their advection of fragmentary sea ice, and the cyclonic atmospheric vortex to the northeast is exposed by its pattern in a stratus cloud deck. Each vortex type probably developed from an antecedent horizontal shear flow in its respective medium.

Vortices are created by a nonlinear advective process of *self-organization*, from an incoherent flow pattern into a coherent one, more local than global. The antecedent conditions for vortex emergence, when it occurs, can either be incoherent forcing and initial conditions or be a late-stage outcome of the *instability* of a prevailing shear flow, from which fluctuations extract energy and thereby amplify. This self-organizing behavior conspicuously contrasts with the nonlinear advective dynamics of *turbulence*. On average turbulence acts to change the flow patterns, to increase their complexity (i.e., their incoherence), and to limit the time over which the evolution is predictable. A central problem in GFD is how these contrasting paradigms – coherent structures and turbulence – can each

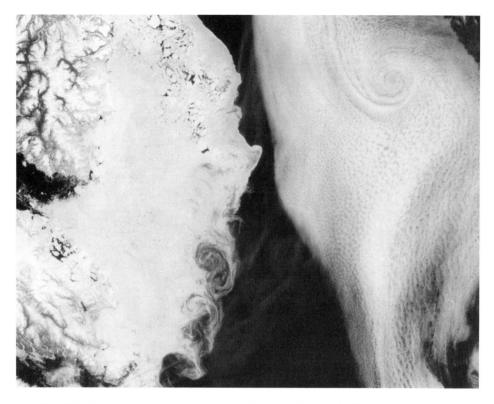

Fig. 3.1. Oceanic and atmospheric vortices in the Davis Strait (north of the Labrador Sea, west of Greenland) during June 2002. Both vortex types are mesoscale vortices with horizontal diameters of tens–hundreds km. (Courtesy of Jacques Descloirest, NASA Goddard Space Flight Center.)

have validity in nature. This chapter is an introduction to these phenomena in the special situation of two-dimensional, or barotropic, fluid dynamics.

3.1 Barotropic equations

Consider *two-dimensional (2D) dynamics*, with $\partial_z = w = \delta\rho = \delta\theta = 0$, and purely vertical rotation with $\mathbf{\Omega} = \hat{\mathbf{z}}\, f/2$. The governing momentum and continuity equations under these conditions are

$$\frac{Du}{Dt} - fv = -\frac{\partial \phi}{\partial x} + F^{(x)}$$

$$\frac{Dv}{Dt} + fu = -\frac{\partial \phi}{\partial y} + F^{(y)}$$

$$\frac{\partial u}{\partial x} + \frac{\partial v}{\partial y} = 0, \qquad\qquad (3.1)$$

with

$$\frac{D}{Dt} = \frac{\partial}{\partial t} + u \frac{\partial}{\partial x} + v \frac{\partial}{\partial y}.$$

These equations conserve the total *kinetic energy*,

$$\mathrm{KE} = \frac{1}{2} \iint dx\, dy\, \mathbf{u}^2, \tag{3.2}$$

when $\mathbf{F}$ is zero and no energy flux occurs through the boundary:

$$\frac{d}{dt}\, \mathrm{KE} = 0 \tag{3.3}$$

(cf., Section 2.1.4 for constant ρ and e in 2D). Equation (3.3) can be derived by multiplying the momentum equation in (3.1) by $\mathbf{u}\cdot$, integrating over the domain, and using continuity to show that there is no net energy source or sink from advection and pressure force. The 2D incompressibility relation implies that the velocity can be represented entirely in terms of a streamfunction, $\psi(x, y, t)$,

$$u = -\frac{\partial \psi}{\partial y}, \quad v = \frac{\partial \psi}{\partial x}, \tag{3.4}$$

since there is no divergence (cf., (2.24)). The vorticity (2.26) in this case only has a vertical component, $\zeta = \zeta^z$:

$$\zeta = \frac{\partial v}{\partial x} - \frac{\partial u}{\partial y} = \nabla^2 \psi. \tag{3.5}$$

(In the present context, it is implicit that $\nabla = \nabla_{\mathrm{h}}$.) There is no buoyancy influence on the dynamics. This is an example of *barotropic flow* using either of its common definitions, $\partial_z = 0$ (sometimes enforced by taking a depth average of a 3D flow) or $\nabla \phi \times \nabla \rho = 0$. (The opposite of barotropic is *baroclinic*; Chapter 5). The consequence of these simplifying assumptions is that the gravitational force plays no overt role in 2D fluid dynamics, however much its influence may be implicit in the rationale for why 2D flows are geophysically relevant (McWilliams, 1983).

3.1.1 Circulation

The circulation (defined in Section 2.1) has a strongly constrained time evolution. This will be shown using an infinitesimal calculus. Consider the time evolution of a line integral $\int_{\mathcal{C}} \mathbf{A} \cdot d\mathbf{r}$, where $\mathbf{A}$ is an arbitrary vector and $\mathcal{C}$ is a closed material curve (i.e., attached to the material parcels along it). A small increment along

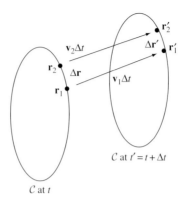

Fig. 3.2. Schematic of circulation evolution for a material line that follows the flow. $\mathcal{C}$ is the closed line at times, t and $t' = t + \Delta t$. The location of two neighboring points are $\mathbf{r}_1$ and $\mathbf{r}_2$ at time t, and they move with velocity $\mathbf{v}_1$ and $\mathbf{v}_2$ to $\mathbf{r}'_1$ and $\mathbf{r}'_2$ at time t'.

the curve between two points marked 1 and 2, $\Delta\mathbf{r} = \mathbf{r}_2 - \mathbf{r}_1$, becomes $\Delta\mathbf{r}'$ after a small interval, Δt (Fig. 3.2):

$$\Delta\mathbf{r}' \equiv \mathbf{r}'_2 - \mathbf{r}'_1$$
$$\approx (\mathbf{r}_2 + \mathbf{u}_2 \Delta t) - (\mathbf{r}_1 + \mathbf{u}_1 \Delta t)$$
$$= \Delta\mathbf{r} + (\mathbf{u}_2 - \mathbf{u}_1)\Delta t, \tag{3.6}$$

using a Taylor series expansion in time for the Lagrangian coordinate, $\mathbf{r}(t)$. Thus,

$$\frac{\Delta\mathbf{r}' - \Delta\mathbf{r}}{\Delta t} \approx \mathbf{u}_2 - \mathbf{u}_1 \approx \frac{\partial\mathbf{u}}{\partial s}\Delta s = (\Delta\mathbf{r}\cdot\nabla)\mathbf{u} \tag{3.7}$$

for small $\Delta s = |\Delta\mathbf{r}|$, where s is arc length along $\mathcal{C}$. As $\Delta t \to 0$, (3.7) becomes

$$\frac{D}{Dt}\Delta\mathbf{r} = (\Delta\mathbf{r}\cdot\nabla)\mathbf{u}. \tag{3.8}$$

This expresses the stretching and bending of $\Delta\mathbf{r}$ through the tangential and normal components of $(\Delta\mathbf{r}\cdot\nabla)\mathbf{u}$, respectively. Now divide $\mathcal{C}$ into small line elements $\Delta\mathbf{r}_i$ to obtain

$$\frac{d}{dt}\int_{\mathcal{C}}\mathbf{A}\cdot d\mathbf{r} \approx \frac{d}{dt}\sum_i \mathbf{A}_i\cdot\Delta\mathbf{r}_i$$

$$= \sum_i\left[\frac{D\mathbf{A}_i}{Dt}\cdot\Delta\mathbf{r}_i + \mathbf{A}_i\cdot\frac{D\Delta\mathbf{r}_i}{Dt}\right]$$

$$= \sum_i\left[\frac{D\mathbf{A}_i}{Dt}\cdot\Delta\mathbf{r}_i + \mathbf{A}_i\cdot(\Delta\mathbf{r}_i\cdot\nabla)\mathbf{u}\right] \tag{3.9}$$

for any vector field, $\mathbf{A}(\mathbf{r}, t)$. The time derivative, d_t, for the material line integral as a whole is replaced by the substantial derivative, D_t, operating on each of the local elements, $\mathbf{A}_i$ and $\Delta \mathbf{r}_i$. As $\max_i |\Delta \mathbf{r}_i| \to 0$, (3.9) becomes

$$\frac{d}{dt} \int_{\mathcal{C}} \mathbf{A} \cdot d\mathbf{r} = \int_{\mathcal{C}} \frac{D\mathbf{A}}{Dt} \cdot d\mathbf{r} + \int_{\mathcal{C}} \mathbf{A} \cdot (d\mathbf{r} \cdot \nabla) \mathbf{u}. \tag{3.10}$$

When $\mathbf{A} = \mathbf{u}$, the last term vanishes because

$$\int_{\mathcal{C}} \mathbf{u} \cdot (d\mathbf{r} \cdot \nabla) \mathbf{u} = \int_{\mathcal{C}} \mathbf{u} \cdot \frac{\partial \mathbf{u}}{\partial s} \, ds$$

$$= \int_{\mathcal{C}} ds \frac{\partial}{\partial s} \left(\frac{1}{2} \mathbf{u}^2 \right)$$

$$= \frac{1}{2} \mathbf{u}^2 |_{\text{start}}^{\text{end}} = 0, \tag{3.11}$$

since the start and end points are the same point for the closed curve, $\mathcal{C}$. Thus,

$$\frac{d}{dt} \int_{\mathcal{C}} \mathbf{u} \cdot d\mathbf{r} = \int_{\mathcal{C}} \frac{D\mathbf{u}}{Dt} \cdot d\mathbf{r}. \tag{3.12}$$

The left-hand side integral operated upon by D_t is called the *circulation*, $\mathcal{C}$. After substituting for the substantial derivative from the momentum equations (3.1),

$$\frac{d}{dt} \int_{\mathcal{C}} \mathbf{u} \cdot d\mathbf{r} = \int_{\mathcal{C}} [-\hat{\mathbf{z}} f \times \mathbf{u} - \nabla \phi + \mathbf{F}] \cdot d\mathbf{r}. \tag{3.13}$$

Two of the right-hand side terms are evaluated as

$$\int_{\mathcal{C}} \nabla \phi \cdot d\mathbf{r} = \int_{\mathcal{C}} \frac{\partial \phi}{\partial s} \, ds = 0, \tag{3.14}$$

and

$$-\int_{\mathcal{C}} \hat{\mathbf{z}} f \times \mathbf{u} \cdot d\mathbf{r} = -\int_{\mathcal{C}} f \mathbf{u} \cdot \hat{\mathbf{n}} \, ds$$

$$= \int_{\mathcal{C}} f \frac{\partial \psi}{\partial s} \, ds$$

$$= -\int_{\mathcal{C}} \psi \frac{\partial f}{\partial s} \, ds, \tag{3.15}$$

again using the fact that the integral of a derivative vanishes. So here the form of *Kelvin's circulation theorem* is

$$\frac{dC}{dt} = \frac{d}{dt} \int_{\mathcal{C}} \mathbf{u} \cdot d\mathbf{r} = \int_{\mathcal{C}} [-\psi \nabla f + \mathbf{F}] \cdot d\mathbf{r}. \tag{3.16}$$

Circulation can only change due to non-conservative viscous or external forces, or due to spatial variation in f, e.g., in the β-plane approximation (Section 2.4). Insofar as the latter are minor effects, as is often true, then circulation is preserved on all material circuits, no matter how much the circuits move around and bend with the advecting flow field. This strongly constrains the evolutionary possibilities for the flow, but the constraint is expressed in an integral, Lagrangian form that is rarely easy to interpret more prosaically.

Baroclinic Kelvin's theorem

As remarked after (3.5), the more fundamental definition of baroclinic is $\nabla p \times \nabla \rho \neq 0$. As a brief diversion, consider Kelvin's theorem for a fully 3D flow to see why $\nabla p \times \nabla \rho$ is germane to non-barotropic dynamics. For a fully compressible fluid, the derivation of Kelvin's circulation theorem includes the following pressure-gradient term on its right-hand side:

$$\frac{dC}{dt} = -\int_{\mathcal{C}} \frac{1}{\rho} \nabla p \cdot d\mathbf{s} + \cdots$$

$$= -\iint_{\mathcal{A}} \hat{\mathbf{n}} \cdot \nabla \times \left[\frac{1}{\rho} \nabla p \right] d\mathcal{A} + \cdots$$

$$= \iint_{\mathcal{A}} \frac{1}{\rho^2} \hat{\mathbf{n}} \cdot \nabla p \times \nabla \rho \, d\mathcal{A} + \cdots . \tag{3.17}$$

The dots denote other contributions not considered here. The gradient operator, ∇, here is fully 3D, $\mathcal{A}$ is the area of the 2D surface interior to the curve $\mathcal{C}$, and $\hat{\mathbf{n}}$ is the unit vector normal to this surface. With the Boussinesq momentum approximation applied to circulation within horizontal planes (i.e., $\hat{\mathbf{n}} = \hat{\mathbf{z}}$),

$$\frac{dC}{dt} = \iint_{\mathcal{A}} \frac{\hat{\mathbf{z}}}{\rho_0^2} \cdot \nabla_{\mathrm{h}} p \times \nabla_{\mathrm{h}} \rho \, dx \, dy + \cdots . \tag{3.18}$$

Therefore, circulation is generated whenever $\nabla_{\mathrm{h}} p \times \nabla_{\mathrm{h}} \rho \neq 0$, the usual situation for 3D, stratified flows. But this will not happen if $\rho = \rho_0$, or $\rho = \bar{\rho}(z)$, or $\{p = \tilde{p}(x, y, t)\hat{p}(z), \ \rho = -(\tilde{p}/g) \, d\hat{p}/dz, \ \mathbf{u}_{\mathrm{h}} = \tilde{\mathbf{u}}_{\mathrm{n}}\hat{p}\}$. The first two of these circumstances are consistent with $\partial \mathbf{u}_{\mathrm{h}}/\partial z = 0$, a 2D flow, while the third one is not. The third circumstance is often called an *equivalent barotropic flow*, whose dynamics have a lot in common with shallow-water flow (Chapter 4). For further discussion see (Gill, 1982, pp. 237–8).

3.1.2 Vorticity and potential vorticity

The *vorticity equation* is derived by taking the curl, $(\hat{\mathbf{z}} \cdot \nabla_{\mathrm{h}} \times)$, of the momentum equations in (3.1). Now examine in turn each term that results from this operation.

An arrow indicates the change in a term coming from the momentum equation after applying the curl:

$$\frac{\partial \mathbf{u}}{\partial t} \longrightarrow -\frac{\partial^2 u}{\partial y \partial t} + \frac{\partial^2 v}{\partial x \partial t} = \frac{\partial \zeta}{\partial t}; \tag{3.19}$$

$$(\mathbf{u} \cdot \nabla)\mathbf{u} \longrightarrow -\frac{\partial}{\partial y}\left[u\frac{\partial u}{\partial x} + v\frac{\partial u}{\partial y}\right] + \frac{\partial}{\partial x}\left[u\frac{\partial v}{\partial x} + v\frac{\partial v}{\partial y}\right]$$

$$= u\left(\frac{\partial^2 v}{\partial x^2} - \frac{\partial^2 u}{\partial y \partial x}\right) + v\left(\frac{\partial^2 v}{\partial y \partial x} - \frac{\partial^2 u}{\partial y^2}\right)$$

$$- \frac{\partial u}{\partial y}\left(\frac{\partial u}{\partial x} + \frac{\partial v}{\partial y}\right) + \frac{\partial v}{\partial x}\left(\frac{\partial u}{\partial x} + \frac{\partial v}{\partial y}\right)$$

$$= \mathbf{u} \cdot \nabla \zeta, \tag{3.20}$$

using the 2D continuity relation in (3.1);

$$\nabla \phi \longrightarrow -\frac{\partial}{\partial y}\left(\frac{\partial \phi}{\partial x}\right) + \frac{\partial}{\partial x}\left(\frac{\partial \phi}{\partial y}\right) = 0; \tag{3.21}$$

$$f\hat{\mathbf{z}} \times \mathbf{u} \longrightarrow -\frac{\partial}{\partial y}(-fv) + \frac{\partial}{\partial x}(fu)$$

$$= f\left(\frac{\partial u}{\partial x} + \frac{\partial v}{\partial y}\right) + u\frac{\partial f}{\partial x} + v\frac{\partial f}{\partial y}$$

$$= \mathbf{u} \cdot \nabla f; \tag{3.22}$$

and

$$\mathbf{F} \longrightarrow -\frac{\partial F^x}{\partial y} + \frac{\partial F^y}{\partial x} = \mathcal{F}. \tag{3.23}$$

The result is

$$\frac{D\zeta}{Dt} = -\mathbf{u} \cdot \nabla f + \mathcal{F}. \tag{3.24}$$

The vorticity only changes following a parcel because of a viscous or external force curl, $\mathcal{F}$, or spatial variation in f. Note the similarity to Kelvin's theorem (3.16). This is to be expected because, as derived in Section 2.1,

$$\int_{\mathcal{C}} \mathbf{u} \cdot d\mathbf{r} = \iint_{\mathcal{A}} \zeta \, dx \, dy. \tag{3.25}$$

Equation (3.24) is a local differential relation, rather than an integral relation, but since (3.16) applies to all possible material curves, both relations cover the entire 2D domain.

Advection operator

Using (3.4) the advection operator can be rewritten as

$$\mathbf{u} \cdot \nabla A = u \frac{\partial A}{\partial x} + v \frac{\partial A}{\partial y}$$

$$= -\frac{\partial \psi}{\partial y} \frac{\partial A}{\partial x} + \frac{\partial \psi}{\partial x} \frac{\partial A}{\partial y} = J[\psi, A] \qquad (3.26)$$

for any advected field, A. J is called the *Jacobian operator*, and it is the approximate form for advection in flows dominated by ψ (cf., Section 2.2.1), even in 3D.

Potential vorticity

Since $\partial_t f = 0$, (3.24) can be rewritten as

$$\frac{Dq}{Dt} = \mathcal{F}, \qquad (3.27)$$

with the *potential vorticity* defined by

$$q = f + \zeta . \qquad (3.28)$$

When $\mathbf{F} = 0$ (conservative flow), q is a parcel invariant; i.e., $D_t q = 0$ for all parcels in the domain. This implies that a conservative flow can only rearrange the spatial distribution of $q(\mathbf{x})$ without changing any of its aggregate (or integral) properties; e.g.,

$$\frac{d}{dt} \iint dx\, dy\, q^n = 0 \qquad (3.29)$$

for any value of n as long as there is no potential-vorticity flux at the boundary, $q\mathbf{u} \cdot \hat{\mathbf{n}} = 0$.

A univariate dynamical system

Equations (3.24) or (3.27), with (3.5) and/or (3.28), comprise a partial differential equation system with ψ as the only dependent variable (assuming that f is known and $\mathcal{F}$ can be expressed in terms of the flow) because ϕ does not appear in the potential vorticity equation, in contrast to the momentum-continuity formulation (3.1). For example, with $f = f_0$ and $\mathbf{F} = 0$, (3.24) can be written entirely in terms of ψ as

$$\nabla^2 \frac{\partial \psi}{\partial t} + J[\psi, \nabla^2 \psi] = 0. \qquad (3.30)$$

Equation (3.30) is often called the *barotropic vorticity equation* since it has no contributions from vertical shear or any other vertical gradients.

Rossby waves

As an alternative to (3.30) when $f = f(y) = f_0 + \beta_0(y - y_0)$ (i.e., the β-plane approximation (2.88)) and advection is neglected (i.e., the flow is linearized about a resting state), (3.24) or (3.28) become

$$\nabla^2 \frac{\partial \psi}{\partial t} + \beta_0 \frac{\partial \psi}{\partial x} = 0. \tag{3.31}$$

In an unbounded domain this equation has normal-mode solutions with eigen-modes,

$$\psi = \text{Real} \left(\psi_0 e^{i(kx + \ell y - \omega t)} \right), \tag{3.32}$$

for an arbitrary amplitude constant, ψ_0 (with the understanding that only the real part of ψ is physically meaningful) and eigenvalues (eigenfrequencies),

$$\omega = -\frac{\beta_0 k}{k^2 + \ell^2}. \tag{3.33}$$

This can be verified as a solution by substitution into (3.31). The type of relation (3.33), between the eigenfrequency and the wavenumbers and environmental parameters (here β_0), is called a dispersion relation, and it is a usual component for wave and instability solutions (Chapter 4). These particular eigenmodes are westward-propagating (i.e., $\omega/k < 0$), barotropic Rossby waves. (Sections 4.6 and 4.7 have more Rossby-wave analyses.)

3.1.3 Divergence and diagnostic force balance

The *divergence equation* is derived by operating on the momentum equations in (3.1) with $(\nabla \cdot)$. Again examine the effect of this operation on each term:

$$\frac{\partial \mathbf{u}}{\partial t} \longrightarrow \frac{\partial}{\partial x}\frac{\partial u}{\partial t} + \frac{\partial}{\partial y}\frac{\partial v}{\partial t} = 0; \tag{3.34}$$

$$(\mathbf{u} \cdot \nabla)\mathbf{u} \longrightarrow \frac{\partial}{\partial x}\left(u\frac{\partial u}{\partial x} + v\frac{\partial u}{\partial y} \right) + \frac{\partial}{\partial y}\left(u\frac{\partial v}{\partial x} + v\frac{\partial v}{\partial y} \right)$$

$$= -2\left(\frac{\partial u}{\partial x}\frac{\partial v}{\partial y} - \frac{\partial u}{\partial y}\frac{\partial v}{\partial x} \right)$$

$$= -2J\left[\frac{\partial \psi}{\partial x}, \frac{\partial \psi}{\partial y} \right]; \tag{3.35}$$

$$-\nabla\phi \longrightarrow -\nabla^2\phi; \tag{3.36}$$

$$f\hat{\mathbf{z}} \times \mathbf{u} \longrightarrow \frac{\partial}{\partial x}(-fv) + \frac{\partial}{\partial y}(fu) = -\nabla \cdot (f\nabla\psi); \tag{3.37}$$

and

$$\mathbf{F} \longrightarrow \mathbf{V} \cdot \mathbf{F}. \tag{3.38}$$

The result is

$$\nabla^2 \phi = \mathbf{V} \cdot (f \mathbf{V} \psi) + 2J \left[\frac{\partial \psi}{\partial x}, \frac{\partial \psi}{\partial y} \right] + \mathbf{V} \cdot \mathbf{F}. \tag{3.39}$$

This relation allows ϕ to be calculated diagnostically from ψ and $\mathbf{F}$, whereas, as explained near (3.30), ψ can be prognostically solved without knowing ϕ. The partial differential equation system (3.27) and (3.39) is fully equivalent to the primitive variable form (3.1), given consistent boundary and initial conditions. The former equation pair is a system that has only a single time derivative. Therefore, it is a first-order system that needs only a single field (e.g., $\psi(\mathbf{x}, 0)$) as the initial condition, whereas an incautious inspection of (3.1), by counting time derivatives, might wrongly conclude that the system is second order, requiring two independent fields as an initial condition. The latter mistake results from overlooking the consequences of the continuity equation that relates the separate time derivatives, $\partial_t u$ and $\partial_t v$. This mistake is avoided for the univariate system because the continuity constraint is implicit in the use of ψ as the prognostic variable.

After neglecting $\mathbf{V} \cdot \mathbf{F}$ and using the following scaling estimates,

$$\mathbf{u} \sim V, \ \mathbf{x} \sim L, \ f \sim f_0,$$

$$\psi \sim VL, \ \phi \sim f_0 VL, \ \beta = \frac{df}{dy} \sim Ro\frac{f_0}{L}, \tag{3.40}$$

for $Ro \ll 1$, (3.39) becomes

$$\nabla^2 \phi = f_0 \nabla^2 \psi [1 + \mathcal{O}(Ro)] \quad \Longrightarrow \quad \phi \approx f_0 \psi. \tag{3.41}$$

This is the *geostrophic balance* relation (2.106). For general f and Ro, the 2D divergence equation is

$$\nabla^2 \phi = \mathbf{V} \cdot (f \mathbf{V} \psi) + 2J \left[\frac{\partial \psi}{\partial x}, \frac{\partial \psi}{\partial y} \right], \tag{3.42}$$

again neglecting $\mathbf{V} \cdot \mathbf{F}$.

This is called the *gradient-wind balance* relation. Equation (3.42) is an exact relation for conservative 2D motions, but is also often a highly accurate approximation for 3D motions with $Ro \leq \mathcal{O}(1)$ and compatible initial conditions and forcing. In comparison geostrophic balance is accurate only if $Ro \ll 1$. When the flow evolution satisfies a diagnostic relation such as (3.41) or (3.42), it is said to have *balanced dynamics* and, by implication, exhibits fewer temporal degrees of freedom than allowed by the more general dynamics. Most large-scale flow evolution is well balanced, but inertial oscillations and internal gravity waves are not balanced (cf., Chapters 2 and 4). Accurate numerical weather forecasts require that the initial conditions of the time integration be well balanced, or the evolution will be erroneously oscillatory compared to nature.

3.1.4 Stationary, inviscid flows

Zonal flow

A parallel flow, such as the zonal flow,

$$\mathbf{u}(\mathbf{x}, t) = U(y)\hat{\mathbf{x}}, \tag{3.43}$$

is a steady flow when $\mathbf{F} = 0$ (e.g., when $\nu = 0$). This is called an inviscid *stationary state*, i.e., a non-evolving solution of (3.1) and (3.28) for which the advection operator is trivial. On the f-plane a stationary parallel flow can have an arbitrary orientation, but on the β-plane, with $f = f(y)$, only a zonal flow (3.43) is a stationary solution. This flow configuration makes the advective potential-vorticity tendency vanish,

$$\frac{\partial q}{\partial t} = -J\,[\psi, q] = -J\left[-\int^y U(y')\,dy',\ f(y) - \frac{dU}{dy}(y)\right] = 0,$$

because the Jacobian operator vanishes if each of its arguments is a function of a single variable, here y. In a zonal flow, all other flow quantities (e.g., ψ, ϕ, ζ, q) are only functions of the coordinate y.

Vortex flow

A simple example of a *vortex solution* for 2D, conservative, uniformly rotating (i.e., $f = f_0$) dynamics is an *axisymmetric flow*, where $\psi(x, y, t) = \psi(r)$, and $r = [(x - x_0)^2 + (y - y_0)^2]^{1/2}$ is the radial distance from the vortex center at (x_0, y_0). This is also a stationary state since in (3.24)–(3.26), $J[\psi(r), \zeta(r)] = 0$ (see (3.75) for the definition of J in cylindrical coordinates). The most common vortex radial shape is a *monopole vortex* (Fig. 3.3). It has a monotonic decay in ψ as r increases away from the extremum at the origin (ignoring a possible far-field behavior of $\psi \propto \log[r]$; see (3.50) below). An axisymmetric solution is most compactly represented in *cylindrical coordinates*, (r, θ), that are related to the Cartesian coordinates, (x, y), by

$$x = x_0 + r\cos\theta, \quad y = y_0 + r\sin\theta. \tag{3.44}$$

The solution has corresponding cylindrical-coordinate velocity components, (U, V), related to the Cartesian components by

$$u = U\cos\theta - V\sin\theta, \quad v = U\sin\theta + V\cos\theta. \tag{3.45}$$

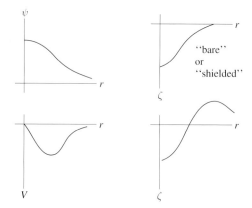

Fig. 3.3. An axisymmetric anticyclonic monopole vortex (when $f_0 > 0$). (Left) typical radial profiles for ψ and V. (Right) typical radial profiles for ζ, showing either a monotonic decay ("bare") or an additional outer annulus of opposite-sign vorticity ("shielded").

Thus, for an axisymmetric vortex,

$$\mathbf{u} = \mathbf{U} = \hat{\mathbf{z}} \times \nabla \psi \longrightarrow V = \frac{\partial \psi}{\partial r}, \quad U = 0$$

$$\zeta = \hat{\mathbf{z}} \cdot \nabla \times \mathbf{u} \longrightarrow \zeta = \frac{1}{r} \frac{\partial}{\partial r}[rV]. \tag{3.46}$$

A compact monopole vortex – whose vorticity, $\zeta(r)$, is restricted to a finite core region (i.e., $\zeta = 0$ for all $r \geq r_*$) – has a nearly universal structure to its velocity in the *far-field* region well away from its center. Integrating the last relation in (3.46) with the boundary condition $V(0) = 0$ (i.e., there can be no azimuthal velocity at the origin where the azimuthal direction is undefined) yields

$$V(r) = \frac{1}{r} \int_0^r \zeta(r')r' \, dr'. \tag{3.47}$$

For $r \geq r_*$, this implies that

$$V(r) = \frac{C}{2\pi r}. \tag{3.48}$$

The associated far-field circulation, C, is

$$C(r) = \int_{r \geq r_*} \mathbf{u} \cdot d\mathbf{r}'$$

$$= \int_0^{2\pi} V(r)r \, d\theta$$

$$= 2\pi r V(r)$$

$$= 2\pi \int_0^{r_*} \zeta(r')r' \, dr'. \tag{3.49}$$

$C(r)$ is independent of r in the far-field; i.e., the vortex has constant circulation around all integration circuits, $\mathcal{C}$, that lie entirely outside r_*.

Also,

$$\frac{\partial \psi}{\partial r} = V$$

$$\Rightarrow \psi = \int_0^r V \, dr' + \psi_0$$

$$\Rightarrow \psi \sim \psi_0 \text{ as } r \to 0$$

$$\Rightarrow \psi \sim \frac{C}{2\pi} \ln r \text{ as } r \to \infty. \qquad (3.50)$$

For monopoles with only a single sign for $\zeta(r)$ (e.g., the "bare" profile in Fig. 3.3), $C \neq 0$. In contrast, for a "shielded" profile (Fig. 3.3), there is a possibility that $C = 0$ due to cancelation between regions with opposite-sign ζ. If $C = 0$, the vortex far-field flow (3.48) is zero to leading order in $1/r$, and $V(r)$ is essentially, though not precisely, confined to the region where $\zeta \neq 0$. In this case its advective influence on its neighborhood is spatially more localized than when $C \neq 0$. Finally, the strain rate for an axisymmetric vortex has the formula,

$$S = r \frac{d}{dr}\left[\frac{V}{r}\right] \sim -\frac{C}{2\pi r^2} \text{ as } r \to \infty. \qquad (3.51)$$

Thus, the strain rate is spatially more extensive than the vorticity for a vortex, but it is less extensive than the velocity field.

Monopole vortices can have either sign for their azimuthal flow direction and the other dynamical variables. Assuming $f_0 > 0$ (northern hemisphere) and geostrophic balance (3.41), the two *vortex parities* are categorized as

cyclonic: $V > 0, \ \zeta > 0, \ C > 0, \ \psi < 0, \ \phi < 0.$

anticyclonic: $V < 0, \ \zeta < 0, \ C < 0, \ \psi > 0, \ \phi > 0.$

(For ζ the sign condition refers to $\zeta(0)$ as representative of the vortex core region.) In the southern hemisphere, cyclonic refers to $V < 0, \ \zeta < 0, \ C < 0, \ \psi > 0$, but still $\phi < 0$, and vice versa for anticyclonic. The 2D dynamical equations for ψ, (3.24) or (3.27) above, are invariant under the following transformation:

$$(\psi, u, v, x, y, t, \mathcal{F}, df/dy) \longleftrightarrow (-\psi, u, -v, x, -y, t, -\mathcal{F}, df/dy), \qquad (3.52)$$

even with $\beta \neq 0$. Therefore, any solution for ψ with one parity implies the existence of another solution with the opposite parity with the direction of motion in y reversed. In general 2D dynamics is *parity invariant*, even though the divergence relation and its associated pressure field are not parity invariant. Specifically, the 2D dynamics of cyclones and anticyclones are essentially equivalent. (This is generally not true for 3D dynamics, except for geostrophic flows; e.g., Section 4.5.)

The more general form of the divergence equation is the gradient-wind balance (3.42). For an axisymmetric state with $\partial_\theta = 0$,

$$\frac{1}{r}\frac{d}{dr}\left[r\frac{\partial\phi}{\partial r}\right] = \frac{f_0}{r}\frac{d}{dr}\left[r\frac{\partial\psi}{\partial r}\right] + \frac{1}{r}\frac{d}{dr}\left[\left(\frac{\partial\psi}{\partial r}\right)^2\right]. \tag{3.53}$$

This can be integrated, $-(\int_r^\infty \cdot r\, dr)$, to obtain

$$\frac{\partial\phi}{\partial r} = fV + \frac{1}{r}V^2. \tag{3.54}$$

This expresses a radial force balance in a vortex among pressure-gradient, Coriolis, and centrifugal forces, respectively. (By induction it indicates that the third term in (3.42) is more generally the divergence of a centrifugal force along a curved, but not necessarily circular, trajectory.) Equation (3.54) is a quadratic algebraic equation for V with solutions,

$$V(r) = -\frac{fr}{2}\left(1\pm\sqrt{1+\frac{4}{f^2 r}\frac{\partial\phi}{\partial r}}\right). \tag{3.55}$$

This solution is graphed in Fig. 3.4. Near the origin (the point marked A),

$$\frac{1}{f^2 r}\frac{\partial\phi}{\partial r}\to 0 \quad (Ro\to 0) \quad \text{and} \quad V\approx\frac{1}{f}\frac{\partial\phi}{\partial r}. \tag{3.56}$$

This relation is the geostrophic balance. At the point marked B,

$$V = -\frac{fr}{2}, \quad \frac{\partial\phi}{\partial r} = -\frac{f^2 r}{4}, \quad \text{and} \quad f+\zeta = 0. \tag{3.57}$$

This corresponds to motionless fluid in a non-rotating (inertial) coordinate frame. Real-valued solutions in (3.55) do not exist for $\partial_r\phi$ values that are more negative than $-f^2 r/4$. If there were such an initial condition, the axisymmetric gradient-wind balance relation could not be satisfied, and the evolution would be such that ∂_t, ∂_θ, and/or ∂_z are non-zero in some combination.

The two solution branches extending to the right from point B correspond to the $\pm$ options in (3.55). The lower branch is dashed as an indication that these solutions are usually centrifugally unstable (Section 3.3.2) and so unlikely to persist.

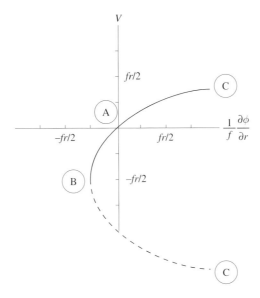

Fig. 3.4. Graphical solution of axisymmetric gradient-wind balance (3.55). The circled point A is the neighborhood of geostrophic balance; the point B is the location of the largest possible negative pressure gradient; and points C are the non-rotating limit $(Ro \to \infty)$ where V can have either sign. Cyclonic and anticyclonic solutions are in the upper and lower half plane, respectively. The dashed line indicates the solution branch that is usually centrifugally unstable for finite Ro values.

Finally, at the points marked C,

$$V \approx \pm\sqrt{r\frac{\partial\phi}{\partial r}} \quad (Ro \to \infty). \tag{3.58}$$

Thus, vortices of either parity must have low-pressure centers (i.e., with $\partial_r\phi > 0$) when rotational influences are negligibly small. This limit for (3.42) and (3.54) is called the *cyclostrophic balance* relation, and it occurs in small-scale vortices with large Ro values (e.g., tornadoes). Property damage from a passing tornado is as much due to the sudden drop of pressure in the vortex core (compared with inside an enclosed building or car) as it is to the drag forces from the extreme wind speed.

Since the gradient-wind balance relation (3.42) is not invariant under the parity transformation (3.52), $\phi(r)$ does not have the the same shape for cyclones and anticyclones when $Ro = \mathcal{O}(1)$ (Fig. 3.5). This disparity is partially the reason why low-pressure minima for cyclonic storms are typically stronger than high-pressure maxima in the extra-tropical atmosphere (though there are also some 3D dynamical reasons for their differences).

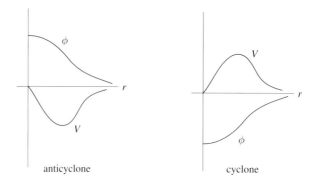

Fig. 3.5. Radial profiles of ϕ and V for axisymmetric cyclones and anticy-clones with finite Rossby number ($f_0 > 0$). Cyclone pressures are "lows," and anticyclone pressures are "highs."

3.2 Vortex movement

A single axisymmetric vortex profile such as (3.46) is a stationary solution when $\nabla f = \mathcal{F} = 0$, and it can be stable to small perturbations for certain profile shapes (Section 3.3). The superposition of several such vortices, however, is not a station-ary solution because axisymmetry is no longer true as a global condition. Multiple vortices induce movement among themselves while more or less preserving their individual shapes as long as they remain well separated from each other; this is because the strain rate is much weaker than the velocity in a vortex far-field. Alternatively, they can cause strong shape changes (i.e., deformations) in each other if they come close enough together.

3.2.1 Point vortices

An idealized model of the mutually induced movement among neighboring vortices is a set of *point vortices*. A point vortex is a singular limit of a stable, axisymmetric vortex with simultaneously $r_* \to 0$ and $\max[\zeta] \to \infty$ while $C \sim \max[\zeta] r_*^2$ is held constant. This limit preserves the far-field information about a vortex in (3.48). The far-field flow is the relevant part for causing mutual motion among well separated vortices. In the point-vortex model the spatial degrees of freedom that represent the shape deformation within a vortex are neglected (but they sometimes do become significantly excited; see Section 3.7).

Mathematical formulae for a point vortex located at $\mathbf{x} = \mathbf{x}_* = (x_*, y_*)$ are

$$\zeta = C\delta(\mathbf{x} - \mathbf{x}_*)$$

$$V = C/2\pi r,$$

$$\psi = c_0 + C/2\pi \ln r$$

$$\mathbf{u} = V\hat{\boldsymbol{\theta}} = V(-\sin\theta\hat{\mathbf{x}} + \cos\theta\hat{\mathbf{y}})$$

$$= V\left(-\frac{y-y_*}{r}\hat{\mathbf{x}} + \frac{x-x_*}{r}\hat{\mathbf{y}}\right)$$

$$= \frac{C}{2\pi r^2}(-(y-y_*)\hat{\mathbf{x}} + (x-x_*)\hat{\mathbf{y}}), \tag{3.59}$$

where $(r, \theta) = \mathbf{x} - \mathbf{x}_*$. We choose $c_0 = 0$ without loss of generality because only the gradient of ψ is related to the velocity. There is a singularity at $r = 0$ for all quantities, and a weak singularity (i.e., logarithmic) at $r = \infty$ for ψ. By superposition, a set of N point vortices located at $\{\mathbf{x}_\alpha, \alpha = 1, N\}$ has the expressions,

$$\zeta(\mathbf{x}, t) = \sum_{\alpha=1}^{N} C_\alpha \delta(\mathbf{x} - \mathbf{x}_\alpha)$$

$$\psi(\mathbf{x}, t) = \frac{1}{2\pi}\sum_{\alpha=1}^{N} C_\alpha \ln|\mathbf{x} - \mathbf{x}_\alpha|$$

$$\mathbf{u}(\mathbf{x}, t) = \frac{1}{2\pi}\sum_{\alpha=1}^{N} \frac{C_\alpha}{|\mathbf{x} - \mathbf{x}_\alpha|^2}[-(y-y_\alpha)\hat{\mathbf{x}} + (x-x_\alpha)\hat{\mathbf{y}})]. \tag{3.60}$$

To show that these fields satisfy the differential relations in (3.46), use the differential relation,

$$\frac{\partial|a|}{\partial a} = \frac{a}{|a|}. \tag{3.61}$$

By (2.1) the trajectory of a fluid parcel is generated from

$$\frac{d\mathbf{x}}{dt}(t) = \mathbf{u}(\mathbf{x}(t), t), \quad \mathbf{x}(0) = \mathbf{x}_0. \tag{3.62}$$

This can be evaluated for any $\mathbf{x}$ using the expression for $\mathbf{u}$ in (3.59). In particular, it can be evaluated for the limit, $\mathbf{x} \to \mathbf{x}_\alpha$, to give

$$\dot{x}_\alpha = -\frac{1}{2\pi}\sum_{\beta}' \frac{C_\beta}{|\mathbf{x}_\alpha - \mathbf{x}_\beta|^2}(y_\alpha - y_\beta)$$

$$\dot{y}_\alpha = +\frac{1}{2\pi}\sum_{\beta}' \frac{C_\beta}{|\mathbf{x}_\alpha - \mathbf{x}_\beta|^2}(x_\alpha - x_\beta), \tag{3.63}$$

with initial conditions, $\mathbf{x}_\alpha(0) = \mathbf{x}_{\alpha 0}$. Here the dot above the variable indicates a time derivative, and the prime denotes a sum over all $\beta \neq \alpha$. This result is based on taking the principal-value limit as $\mathbf{x} \to \mathbf{x}_\alpha$ that gives zero contribution from the right-hand side of $\mathbf{u}$ in (3.60) at the vortex locations, $\mathbf{x}_\alpha$. Equations (3.63) are the equations of motion for the point-vortex system in combination with

$$\dot{C}_\alpha = 0 \;\Rightarrow\; C_\alpha(t) = C_\alpha(0), \tag{3.64}$$

which comes from $D\zeta/Dt = 0$, the conservation of vorticity and circulation following the point-vortex parcels. These equations comprise a well-posed dynamical system that is a temporal ordinary differential equation system of order N. This system can be written even more concisely using complex variables. For the complex vortex position,

$$Z_\alpha = x_\alpha + iy_\alpha, \tag{3.65}$$

(3.63) becomes

$$\dot{Z}_\alpha^* = \frac{1}{2\pi i} \sum_\beta{}' \frac{C_\beta}{Z_\alpha - Z_\beta}. \tag{3.66}$$

The symbol $\cdot^*$ denotes the complex conjugate of a variable.

Yet another way to write (3.63)–(3.66) is as a *Hamiltonian dynamical system*. Transformed dependent variables are defined by

$$p_\alpha = C_\alpha^{1/2} x_\alpha, \quad q_\alpha = C_\alpha^{1/2} y_\alpha. \tag{3.67}$$

Using these variables, the equations of motion (3.63) are written in "canonical form," namely

$$\dot{p}_\alpha = \frac{\partial H}{\partial q_\alpha}, \quad \dot{q}_\alpha = -\frac{\partial H}{\partial p_\alpha}, \tag{3.68}$$

with the Hamiltonian functional for point vortices defined by

$$H(p_\alpha, q_\alpha) = -\frac{1}{2\pi} \sum_{\alpha,\beta}{}' C_\alpha C_\beta \ln|\mathbf{x}_\alpha - \mathbf{x}_\beta|. \tag{3.69}$$

The primed summation again excludes all terms with $\alpha = \beta$. As is often true in Hamiltonian mechanics, H is interpreted as the energy of the system. In 2D dynamics the only type of energy is the kinetic energy, KE (3.2). In this context H is called the *interaction kinetic energy* since it does not include the *self-energy* contributions to KE associated with the quadratic product of the internal recirculating flow within each vortex. For point vortices, the internal recirculation is both time-invariant for each vortex, hence irrelevant to the dynamics of vortex motion, and infinite in magnitude, due to the spatial singularity in (3.60). H is an integral invariant of the dynamics:

$$
\begin{aligned}
\dot{H} &= \sum_\alpha \left[\frac{\partial H}{\partial p_\alpha} \dot{p}_\alpha + \frac{\partial H}{\partial q_\alpha} \dot{q}_\alpha \right] \\
&= \sum_\alpha \left[\frac{\partial H}{\partial p_\alpha} \left(\frac{\partial H}{\partial q_\alpha} \right) + \frac{\partial H}{\partial q_\alpha} \left(-\frac{\partial H}{\partial p_\alpha} \right) \right] = 0.
\end{aligned}
\tag{3.70}
$$

The time invariance of H is an expression of *energy conservation* in the point-vortex approximation to the general expression for 2D fluid dynamics (3.3). Other integral invariants of (3.63)–(3.69) are

$$
\mathcal{X} \equiv \sum_\alpha C_\alpha x_\alpha / \sum_\alpha C_\alpha
$$

$$
\mathcal{Y} \equiv \sum_\alpha C_\alpha y_\alpha / \sum_\alpha C_\alpha
$$

$$
\mathcal{I} \equiv \sum_\alpha C_\alpha (x_\alpha^2 + y_\alpha^2).
\tag{3.71}
$$

This can be verified by taking the time derivative and substituting the equations of motion. These quantities are point-vortex counterparts of the *vorticity centroid*, $\iint \mathbf{x} \zeta \, d\mathbf{x}$, and *angular momentum*, $\iint r^2 \zeta \, d\mathbf{x}$, integral invariants of conservative 2D dynamics (cf., (3.83) below). They arise – as is usual for conservation laws in Hamiltonian dynamics – because of the invariance (i.e., symmetry) of the Hamiltonian functional, H, with respect to translation and rotation (i.e., spatial location and orientation angle). For example, with the definition of a translated coordinate, $\mathbf{x}' = \mathbf{x} + \mathbf{d}$, for constant $\mathbf{d}$, the form of $H(\mathbf{x}')$ in (3.69) is unchanged from $H(\mathbf{x})$ since $|\mathbf{x}'_\alpha - \mathbf{x}'_\beta| = |\mathbf{x}_\alpha - \mathbf{x}_\beta|$. Thus, there are 4 integrals of the motion and $2N$ degrees of freedom (i.e., the (x_α, y_α)) in the dynamical system, hence $2N - 4$ independent degrees of freedom in a point-vortex system.

There are infinitely more integral invariants of conservative, 2D fluid dynamics that are the counterparts of $\iint \mathcal{G}[q] \, d\mathbf{x}$ for arbitrary functionals $\mathcal{G}$ and $f = f(y)$; cf., (3.29). For point vortices with $f = f_0$, however, these invariants are redundant with the relations, $\dot{C}_\alpha = 0$ for all α, and thus they do not further constrain the independent degrees of freedom for the flow evolution.

There is a well developed theory of Hamiltonian dynamical systems that is therefore applicable to point-vortex dynamics and therefore, approximately, to geophysical fluid dynamics. (See Salmon (1998) for an extensive treatment.) Based upon the independent degrees of freedom, it can be demonstrated that for $N \leq 3$, the point-vortex dynamics are *integrable*. This implies certain limits on the possible complexity of their trajectories. The most complex possibility is *quasi-periodic* trajectories (i.e., expressible as periodic functions in time, with possibly multiple frequencies as, e.g., in $a\cos[\omega_1 t] + b\sin[\omega_2 t]$). In contrast, for $N \geq 4$, point-vortex solutions are not generally integrable, and they commonly have *chaotic* trajectories (Fig. 3.6). (The particular value of N for the onset of chaos depends on the boundary conditions; here an unbounded domain is assumed.)

Now analyze some simple examples of point-vortex solutions. If $N = 1$, the vortex is purely stationary, reflecting the fact that any axisymmetric flow is a stationary state (Section 3.1.4). If $N = 2$, the solution is not stationary, but must be

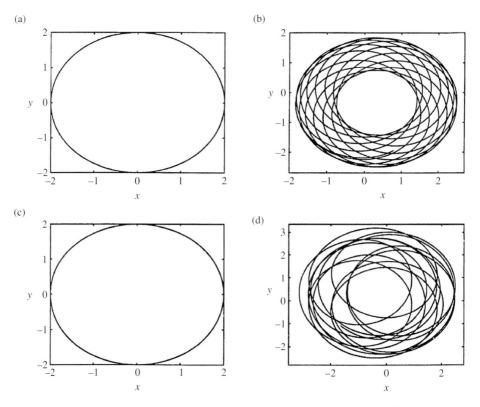

Fig. 3.6. Trajectories for an individual point vortex initially at $(-\sqrt{3}, -1)$. All vortices in the associated point-vortex system are of equal strength with $C > 0$. Initial conditions are the following: (a) three vortices symmetrically located at $(\pm\sqrt{3}, -1)$, $(0, 2)$; (b) three vortices asymmetrically located at $(\pm\sqrt{3}, -1)$, $(1, 1)$; (c) four vortices symmetrically located at $(\pm\sqrt{3}, -1)$, $(0, 2)$, $(0, 0)$; and (d) four vortices asymmetrically located at $(\pm\sqrt{3}, -1)$, $(1, 1)$, $(-2, 2.4)$. The motion is periodic in time for (a)–(c), but it is chaotic for (d). The trajectories in panels (a)–(c) are time periodic, although the period for (c) includes many circuits. The trajectory in (d) is chaotic, and if it were continued for a longer time, the mesh of lines would become much denser.

either a periodic rotation (if $C_1 + C_2 \neq 0$) or else steady propagation (if $C_1 = -C_2$, a *dipole vortex*).

Example 1 Two vortices of the same parity with $C_1 \geq C_2 > 0$. In this case the motion is counter-clockwise rotation around a point lying along the line between the vortex centers (Fig. 3.7, top).

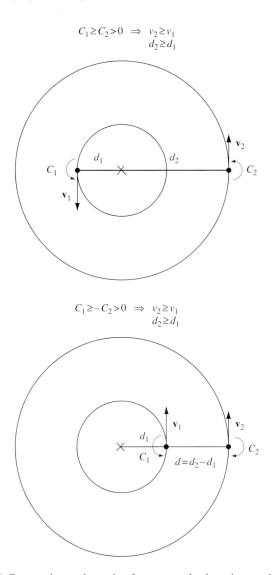

Fig. 3.7. (Top) Co-rotating trajectories for two cyclonic point vortices of unequal circulation strength. (Bottom) Trajectories for a cyclonic and anticyclonic pair of point vortices of unequal circulation strength. Vortex 1 has stronger circulation magnitude than vortex 2. $\times$ denotes the center of rotation for the trajectories, and d_1 and d_2 are the distances from it to the two vortices. The vortex separation is $d = d_1 + d_2$ above and $d = d_2 - d_1$ below.

Example 2 Two vortices of opposite parity with $C_1 \geq -C_2 > 0$. In this case the motion is counter-clockwise rotation around a point located along the line between the vortex centers but on the other side of vortex 1 than the side nearer vortex 2. As $C_2 \to -C_1$, that point of rotation moves to ∞, and the motion becomes a uniform translation perpendicular to the line between the centers (Fig. 3.7, bottom).

Example 3 One vortex near a straight boundary (Fig. 3.8). The boundary condition of $\mathbf{u} \cdot \hat{\mathbf{n}} = 0$ for a bounded half-plane domain can be satisfied by superimposing the velocity fields of two vortices in an infinite domain. One vortex is the real one, and the other is an *image vortex* of the opposite sign at an equal distance located on the other side of the boundary. Thus, the real vortex moves parallel to the boundary at the same speed as this dipole vortex would in an unbounded fluid.

Example 4 A *vortex street* of uniformly spaced vortices, equal in both parity and strength (with, e.g., $C > 0$), lying along an infinite line (Fig. 3.9). This is a stationary configuration since for each vortex the velocity due to any other vortex on its right is canceled by the velocity from the equidistant vortex on the left. However, it is unstable to a *vortex pairing instability*: if two adjacent vortices are perturbed to lie slightly closer together or off the line of centers, they will tend to move away from the line or closer together; this can be verified by calculating the resultant motions in the perturbed configuration. This is a positive feedback (i.e., instability) in the sense that infinitesimal perturbations will continue to grow to finite displacements. In the limit of vanishing vortex separation, the vortex street becomes a *vortex sheet*, representing a flow with a velocity discontinuity across the line; i.e., there is infinite horizontal shear and vorticity at the sheet. Thus, such

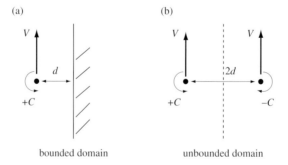

Fig. 3.8. (a) Trajectory of a cyclonic vortex with circulation, $+C$, located a distance, d, from a straight, free-slip boundary and (b) its equivalent image vortex system in an unbounded domain that has zero normal velocity at the location of the virtual boundary. The vortex movement is poleward parallel to the boundary at a speed, $V = C/4\pi d$.

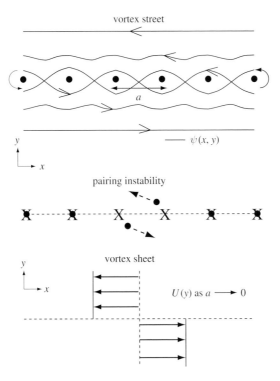

Fig. 3.9. (Top) a vortex street of identical cyclonic point vortices (black dots) lying on a line, with a uniform pair separation distance, a. This is a stationary state since the advective effect of every neighboring vortex is canceled by the opposite effect from the neighbor on the other side. The associated streamfunction contours are shown with arrows indicating the flow direction. (Middle) the instability mode for a vortex street that occurs when two neighboring vortices are displaced to be closer to each other than a, after which they move away from the line and even closer together. "X" denotes the unperturbed street locations. (Bottom) the discontinuous zonal flow profile, $\mathbf{u} = U(y)b\hat{f}x$, of a vortex sheet. This is the limiting flow for a street when $a \to 0$ (or, equivalently, when the flow is sampled a distance away from the sheet much larger than a).

a shear flow is unstable at vanishingly small perturbation length scales (due to the infinitesimal width of the shear layer). This is an example of barotropic instability (Section 3.3) that sometimes is called *Kelvin–Helmholtz instability*. A linear, normal-mode instability analysis for a vortex sheet is presented in Section 3.3.3.

Example 5 A *Karman vortex street* (named after Theodore von Karman). This is a double vortex street of vortices of equal strengths, opposite parities, and staggered positions (Fig. 3.10). Each of the vortices moves steadily along its own row with speed U. This configuration can be shown to be stable to small perturbations if $\cosh[b\pi/a] = \sqrt{2}$, with a the along-line vortex separation and b the between-row

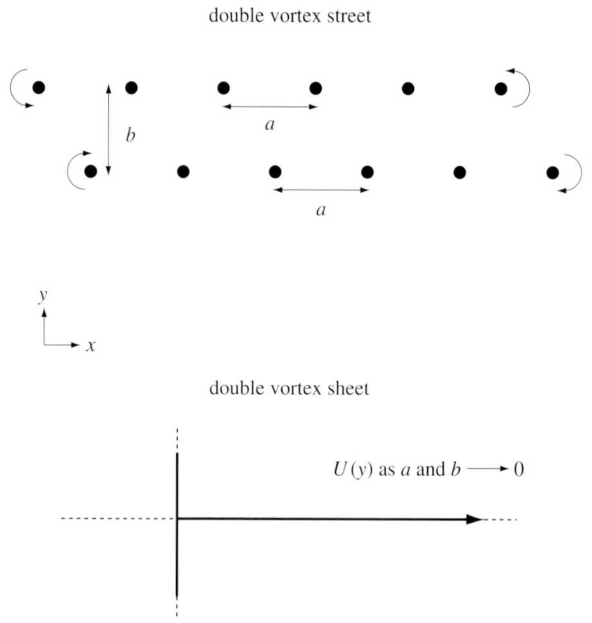

double vortex street

double vortex sheet

Fig. 3.10. (Top) a double vortex street (sometimes called a Karman vortex street) with identical cyclonic vortices on the upper line and identical anticyclonic vortices on the parallel lower line. The vortices (black dots) are separated by a distance, a, along the lines, the lines are separated by a distance, b, and the vortex positions are staggered between the lines. This is a stationary state that is stable to small displacements if $\cosh[\pi b/a] = \sqrt{2}$. (Bottom) as the vortex separation distances shrink to zero, the flow approaches an infinitely thin zonal jet. This is sometimes called a double vortex sheet.

separation. Such a configuration often arises from flow past an obstacle (e.g., a mountain or an island). As $a, b \to 0$, this configuration approaches an infinitely thin jet flow. Alternatively it could be viewed as a double vortex sheet. A finite-separation vortex street is stable, while a finite-width jet is unstable (Section 3.3), indicating that the limit of vanishing separation and width is a delicate one.

3.2.2 Chaos and limits of predictability

An important property of chaotic dynamics is the *sensitive dependence* of the solution to perturbations: a microscopic difference in the initial vortex positions leads to a macroscopic difference in the vortex configuration at a later time of the order of the advection time scale, $T = L/V$. This is the essential reason why the *predictability* of the weather is only possible for a finite time (at most 15–20 d), no matter how accurate the prediction model.

Insofar as chaotic dynamics thoroughly entangles the trajectories of the vortices, then all neighboring, initially well separated parcels will come arbitrarily close together at some later time. This process is called *stirring*. The tracer concentrations carried by the parcels may therefore mix together if there is even a very small tracer diffusivity in the fluid. *Mixing* is blending by averaging the tracer concentrations of separate parcels, and it has the effect of diminishing tracer variations. Trajectories do not mix, because Hamiltonian dynamics is *time reversible*, and any set of vortex trajectories that begins from an orderly configuration, no matter how entangled later, can always be disentangled by reversing the sign of C_α, hence of $\mathbf{u}_\alpha$, and integrating forward over an equivalent time since the initialization. (This is equivalent to reversing the sign of t, while keeping the same sign for the C_α.) Thus, conservative chaotic dynamics *stirs* parcels but *mixes* a passive tracer field with non-zero diffusivity. Equation (3.60) says that non-vortex parcels are also advected by the vortex motion and therefore also stirred, though the stirring efficiency is weak for parcels far away from all vortices. Trajectories of non-vortex parcels can be chaotic even for $N = 3$ vortices in an unbounded 2D domain.

3.3 Barotropic and centrifugal instability

Stationary flows may or may not be stable with respect to small perturbations (cf., Section 2.3.3). This possibility is analyzed here for several types of 2D flow.

3.3.1 Rayleigh's criterion for vortex stability

An analysis is first made for the linear, normal-mode *stability* of a stationary, axisymmetric vortex, $(\overline{\psi}(r),\ \overline{V}(r),\ \overline{\zeta}(r))$ with $f = f_0$ and $\mathcal{F} = 0$ (Section 3.1.4). Assume that there is a small-amplitude streamfunction perturbation, ψ', such that

$$\psi = \overline{\psi}(r) + \psi'(r, \theta, t), \tag{3.72}$$

with $\psi' \ll \overline{\psi}$. Introducing (3.72) into (3.24) and linearizing around the stationary flow (i.e., neglecting terms of $\mathcal{O}(\psi'^2)$ because they are small) yields

$$\nabla^2 \frac{\partial \psi'}{\partial t} + J[\overline{\psi}, \nabla^2 \psi'] + J[\psi', \nabla^2 \overline{\psi}] \approx 0, \tag{3.73}$$

or, recognizing that $\overline{\psi}$ depends only on r,

$$\nabla^2 \frac{\partial \psi'}{\partial t} + \frac{1}{r} \frac{\partial \overline{\psi}}{\partial r} \nabla^2 \frac{\partial \psi'}{\partial \theta} - \frac{1}{r} \frac{\partial \overline{\zeta}}{\partial r} \frac{\partial \psi'}{\partial \theta} \approx 0. \tag{3.74}$$

These expressions use the definitions of the cylindrical-coordinate operators,

$$J[A, B] \equiv \frac{1}{r}\left[\frac{\partial A}{\partial r}\frac{\partial B}{\partial \theta} - \frac{\partial A}{\partial \theta}\frac{\partial B}{\partial r}\right]$$

$$\nabla^2 A \equiv \frac{1}{r}\frac{\partial}{\partial r}\left[r\frac{\partial A}{\partial r}\right] + \frac{1}{r^2}\frac{\partial^2 A}{\partial \theta^2}. \tag{3.75}$$

Now seek *normal mode* solutions to (3.74) with the following space-time structure:

$$\psi'(r, \theta, t) = \text{Real}\,[g(r)e^{i(m\theta - \omega t)}]$$

$$= \frac{1}{2}[g(r)e^{i(m\theta - \omega t)} + g^*(r)e^{-i(m\theta - \omega^* t)}]. \tag{3.76}$$

Inserting (3.76) into (3.74) and factoring out $\exp[i(m\theta - \omega t)]$ leads to the following relation:

$$\frac{1}{r}\partial_r[r\partial_r g] - \frac{m^2}{r^2}g = -\frac{\partial_r \overline{\zeta}}{\frac{\omega r}{m} - \partial_r \overline{\psi}}\,g. \tag{3.77}$$

Next operate on this equation by $\int_0^\infty rg^* \cdot dr$, noting that

$$\int_0^\infty g^*\partial_r[r\partial_r g]\,dr = -\int_0^\infty r(\partial_r g^*)\,(r\partial_r g)\,dr$$

if g or $\partial_r g = 0$ at $r = 0, \infty$ (NB, these are the appropriate boundary conditions for this eigenmode problem). Also, recall that $aa^* = |a|^2 \geq 0$. After integrating the first term in (3.77) by parts, the result is

$$\int_0^\infty r\left[|\partial_r g|^2 + \frac{m^2}{r^2}|g|^2\right]dr = \int_0^\infty \left[\frac{\partial_r \overline{\zeta}}{\frac{\omega}{m} - \frac{1}{r}\partial_r \overline{\psi}}\right]|g|^2\,dr. \tag{3.78}$$

The left-hand side is always real. After writing the complex eigenfrequency as

$$\omega = \gamma + i\sigma \tag{3.79}$$

(i.e., admitting the possibility of perturbations growing at an exponential rate, $\psi' \propto e^{\sigma t}$, called a normal-mode instability), then the imaginary part of the preceding equation is

$$\sigma m \int_0^\infty \left[\frac{\partial_r \overline{\zeta}}{(\gamma - \frac{m}{r}\partial_r \overline{\psi})^2 + \sigma^2}\right]|g|^2\,dr = 0. \tag{3.80}$$

If $\sigma, m \neq 0$, then the integral must vanish. But all terms in the integrand are non-negative except $\partial_r \overline{\zeta}$. Therefore, a necessary condition for instability is that $\partial_r \overline{\zeta}$ must change sign for at least one value of r so that the integrand can have both positive and negative contributions that cancel each other. This is called the Rayleigh inflection point criterion (since the point in r where $\partial_r \overline{\zeta} = 0$ is an inflection point for the vorticity profile, $\overline{\zeta}(r)$). This type of instability is called *barotropic instability* since it arises from horizontal shear and the unstable perturbation flow can lie entirely within the plane of the shear (i.e., comprise a 2D flow).

With reference to the vortex profiles in Fig. 3.3, a bare monopole vortex with monotonic $\overline{\zeta}(r)$ is stable by the Rayleigh criterion, but a shielded vortex may be unstable. More often than not for barotropic dynamics with large Re, what may be unstable is unstable.

3.3.2 Centrifugal instability

There is another type of instability that can occur for a barotropic axisymmetric vortex with constant f. It is different from the one in the preceding section in two important ways. It can occur with perturbations that are uniform along the mean flow, i.e., with $m = 0$; hence it is sometimes referred to as *symmetric instability* even though it can also occur with $m \neq 0$. And the flow field of the unstable perturbation has non-zero vertical velocity and vertical variation, unlike the purely horizontal velocity and structure in (3.76). Its other common names are *inertial instability* and *centrifugal instability*. The simplest way to demonstrate this type of instability is by a parcel displacement argument analogous to that for buoyancy oscillations and convection (Section 2.3.3). Assume there exists an axisymmetric barotropic mean state, $(\partial_r \overline{\phi}, \overline{V}(r))$, that satisfies the gradient-wind balance (3.54). Expressed in cylindrical coordinates, parcels displaced from their mean position, r_o, to $r_o + \delta r$ experience a radial acceleration given by the radial momentum equation,

$$\frac{DU}{Dt} = \frac{D^2 \delta r}{Dt^2} = \left[-\frac{\partial \phi}{\partial r} + fV + \frac{V^2}{r} \right]_{r=r_o+\delta r}. \qquad (3.81)$$

The terms on the right-hand side are evaluated by two principles:

- instantaneous adjustment of the parcel pressure gradient to the local value,

$$\frac{\partial \phi}{\partial r}(r_o + \delta r) = \frac{\partial \overline{\phi}}{\partial r}(r_o + \delta r); \qquad (3.82)$$

- parcel conservation of absolute angular momentum for axisymmetric flow (cf., Section 4.3),

$$A(r_o + \delta r) = \overline{A}(r_o), \qquad A(r) = \frac{fr^2}{2} + rV(r). \qquad (3.83)$$

By using these relations to evaluate the right-hand side of (3.81) and making a Taylor series expansion to express all quantities in terms of their values at $r = r_o$ through $\mathcal{O}(\delta r)$ (cf., (2.69)), the following equation is derived:

$$\frac{D^2 \, \delta r}{Dt^2} + \gamma^2 \, \delta r = 0, \tag{3.84}$$

where

$$\gamma^2 = \frac{1}{2r^3} \overline{A} \frac{d\overline{A}}{dr}\bigg|_{r=r_o}. \tag{3.85}$$

The angular momentum gradient is

$$\frac{dA}{dr} = r\left[f + \frac{1}{r}\frac{d}{dr}[rV] \right]; \tag{3.86}$$

i.e., it is proportional to the absolute vorticity, $f + \zeta$. Therefore, if γ^2 is positive everywhere in the domain (as it is certain to be for approximately geostrophic vortices near point A in Fig. 3.4), the axisymmetric parcel motion will be oscillatory in time around $r = r_o$. However, if $\gamma^2 < 0$ anywhere in the vortex, then parcel displacements in that region can exhibit exponential growth; i.e., the vortex is unstable. At point B in Fig. 3.4, $\overline{A} = 0$, hence $\gamma^2 = 0$. This is therefore a possible marginal point for centrifugal instability. When centrifugal instability occurs, it involves vertical motions as well as the horizontal ones that are the primary focus of this chapter.

3.3.3 Barotropic instability of parallel flows

Free shear layer

Lord Kelvin (as he is customarily called in the GFD community) made a pioneering calculation in the nineteenth century of the unstable 2D eigenmodes for a vortex sheet (cf., the point-vortex street; Section 3.2.1, example 4) located at $y = 0$ in an unbounded domain, with equal and opposite mean zonal flows of $\pm U/2$ on either side. This step-function velocity profile is the limiting form for a continuous profile with

$$\begin{aligned}
\overline{u}(y) &= \frac{Uy}{D}, \quad |y| \leq \frac{D}{2}, \\
&= +\frac{U}{2}, \quad y > \frac{D}{2}, \\
&= -\frac{U}{2}, \quad y < -\frac{D}{2},
\end{aligned} \tag{3.87}$$

as D, the width of the shear layer, vanishes. Such a zonal flow is a stationary state (Section 3.1.4). A mean flow with a one-signed velocity change away from any boundaries is also called a *free shear layer* or a *mixing layer*. The latter term emphasizes the turbulence that develops after the growth of the linear instability that is sometimes called a Kelvin–Helmholtz instability, to a finite-amplitude state where the linearized, normal-mode dynamics are no longer valid (Section 3.6). Because the mean flow has uniform vorticity (zero outside the shear layer and $-U/D$ inside) the perturbation vorticity must be zero in each of these regions since all parcels must conserve their potential vorticity, hence also their vorticity when $f = f_0$. Analogous to the normal modes with exponential solution forms in (3.32) and (3.76), the unstable modes here have a space-time structure (eigensolution) of the form,

$$\psi' = \text{Real}\left(\Psi(y)\, e^{ikx + st}\right). \tag{3.88}$$

k is the zonal wavenumber, and s is the unstable growth rate when its real part is positive. Since $\nabla^2 \psi' = 0$, the meridional structure is a linear combination of exponential functions of ky consistent with perturbation decay as $|y| \to \infty$ and continuity of ψ' at $y = \pm D/2$, namely,

$$\Psi(y) = \Psi_+\, e^{-k(y - D/2)}, \quad y \geq D/2,$$
$$= \left(\frac{\Psi_+ + \Psi_-}{2}\right)\frac{\cosh[ky]}{\cosh[kD/2]} + \left(\frac{\Psi_+ - \Psi_-}{2}\right)\frac{\sinh[ky]}{\sinh[kD/2]},$$
$$- D/2 \leq y \leq D/2,$$
$$= \Psi_-\, e^{k(y + D/2)}, \quad y \leq -D/2. \tag{3.89}$$

The constants, Ψ_+ and Ψ_-, are determined from continuity of both the perturbation pressure, ϕ', and the linearized zonal momentum balance,

$$\frac{\partial u'}{\partial t} + \bar{u}\frac{\partial u'}{\partial x} - \left(f - \frac{\partial \bar{u}}{\partial y}\right)v' = -\frac{\partial \phi'}{\partial x}, \tag{3.90}$$

across the layer boundaries at $y = \pm D/2$, with u', v' evaluated in terms of ψ' from (3.88) and (3.89). These matching conditions yield an eigenvalue equation:

$$s^2 = \left(\frac{kU}{2}\right)^2 \left(2\frac{1 + (1 - [kD]^{-1})\tanh[kD]}{kD(1 + [2]^{-1}\tanh[kD])} - 1\right). \tag{3.91}$$

In the vortex-sheet limit (i.e., $kD \to 0$), there is an instability with $s \to \pm kU/2$. Its growth rate increases as the perturbation wavenumber increases up to a scale comparable to the inverse layer thickness, $1/D \to \infty$. Since s has a zero imaginary part, this instability is a standing mode that amplifies in place without propagation along the mean flow. The instability behavior is consistent with the pairing instability of the finite vortex street approximation to a vortex sheet (Section 3.2.1, Example 4). On the other hand, for very small-scale perturbations with $kD \to \infty$, (3.91) implies that $s^2 \to -(kU/2)^2$; i.e., the eigenmodes are stable and zonally propagating in either direction.

Bickley Jet

In nature shear is spatially distributed rather than singularly confined to a vortex sheet. A well studied example of a stationary zonal flow (Section 3.1.4) with distributed shear is the so-called Bickley Jet,

$$U(y) = U_0 \, \mathrm{sech}^2[y/L_0] = \frac{U_0}{\cosh^2[y/L_0]}, \qquad (3.92)$$

in an unbounded domain. This flow has its maximum speed at $y = 0$ and decays exponentially as $y \to \pm\infty$ (Fig. 3.11). From (3.27) the linearized, conservative, f-plane, potential-vorticity equation for perturbations, ψ, is

$$\left(\frac{\partial}{\partial t} + U\frac{\partial}{\partial x}\right)\nabla^2\psi - \frac{d^2 U}{dy^2}\frac{\partial \psi}{\partial x} = 0. \qquad (3.93)$$

Analogous to Section 3.3.1, a Rayleigh necessary condition for the instability of a parallel flow can be derived for normal-mode eigensolutions of the form,

$$\psi = \mathrm{Real}\left(\Psi(y)e^{ik(x-ct)}\right), \qquad (3.94)$$

with the result that $\partial_y \overline{q} = -\partial_y^2 U$ has to be zero somewhere in the domain. For the Bickley Jet this condition is satisfied because there are two inflection points

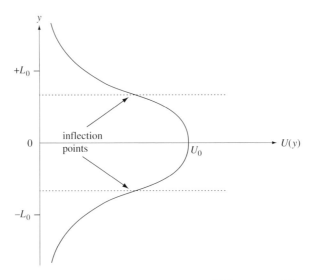

Fig. 3.11. Bickley Jet zonal flow profile, $\mathbf{u} = U(y)\hat{\mathbf{x}}$, with $U(y)$ from (3.92). Inflection points, where $U_{yy} = 0$, occur on the flanks of the jet.

located at $y = \pm 0.66 L_0$. With the β-plane approximation, the Rayleigh criterion for a zonal flow is that

$$\frac{d\bar{q}}{dy} = \beta_0 - \frac{d^2 U}{dy^2} = 0$$

somewhere in the flow. Thus, for a given shear flow, $U(y)$, with inflection points, $\beta \neq 0$ usually has a stabilizing influence (cf., Section 5.2.1 for an analogous β effect for baroclinic instability).

The eigenvalue problem that comes from substituting (3.94) into (3.93) is the following:

$$\Psi_{yy} - \left(k^2 + \frac{\partial_y^2 U}{U - c} \right) \Psi = 0, \quad |\Psi| \to 0 \text{ as } |y| \to \infty. \quad (3.95)$$

This problem, as most shear-flow instability problems, cannot be solved analytically. But it is rather easy to solve numerically as a 1D boundary-value problem as long as there is no singularity in the coefficient in (3.95) associated with a *critical layer* at the y location where $U(y) = c$. Since the imaginary part, c^{im}, of $c = c^{\mathrm{r}} + ic^{\mathrm{im}}$ is non-zero for unstable modes and since, therefore,

$$\frac{1}{U - c} = \frac{(U - c^{\mathrm{r}}) + ic^{\mathrm{im}}}{(U - c^{\mathrm{r}})^2 + (c^{\mathrm{im}})^2}$$

is bounded for all y, these modes do not have critical layers and are easily calculated numerically.

Results are shown in Fig. 3.12. There are two types of unstable modes, a more rapidly growing one where Ψ is an even function in y (i.e., a varicose mode with perturbed streamlines that bulge and contract about $y = 0$, while propagating in x with phase speed, c^{r}, and amplifying with growth rate, $kc^{\mathrm{im}} > 0$) and another one where Ψ is an odd function (i.e., a sinuous mode with streamlines that meander in y) that also propagates in x and amplifies. The unstable growth rates are a modest fraction of the advective rate for the mean jet, U_0/L_0. Both modal types are unstable for all long-wave perturbations with $k < k_{\mathrm{cr}}$, but the value of the critical wavenumber, $k_{\mathrm{cr}} = \mathcal{O}(1/L_0)$, is different for the two modes. Both mode types propagate in the direction of the mean flow with a phase speed $c^{\mathrm{r}} = \mathcal{O}(U_0)$. The varicose mode grows more slowly than the sinuous mode for any specific k. These unstable modes are not consistent with the stable double vortex street (Section 3.2.1, Example 5) as the vortex spacing vanishes, indicating that both stable and unstable behaviors may occur in a given situation.

When viscosity effects are included for a Bickley Jet (overlooking the fact that (3.92) is no longer a stationary state of the governing equations), then the instability is weakened due to the general damping effect of molecular diffusion on the flow, and it can even be eliminated at large enough ν, hence small

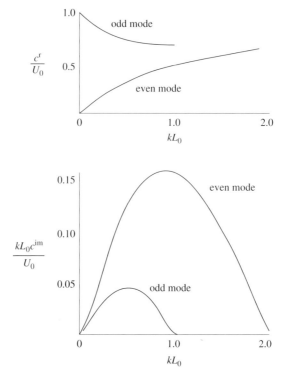

Fig. 3.12. Eigenvalues for the barotropic instability of the Bickley Jet: (a) the real part of the zonal phase speed, c^r, and (b) the growth rate, kc^{im}. (Drazin & Reid, 2004, Fig. 4.25.)

enough Re. Viscosity can also contribute to removing critical-layer singularities among the otherwise stable eigenmodes by providing c with a negative imaginary part, $c^{im} < 0$.

For more extensive discussions of these and other 2D and 3D shear instabilities, see Drazin & Reid (2004).

3.4 Eddy–mean interaction

A normal-mode instability, such as barotropic instability, demonstrates how the amplitude of a perturbation flow can grow with time. Because kinetic energy, KE, is conserved when $\mathbf{F} = 0$ (3.3) and KE is a quadratic functional of $\mathbf{u} = \overline{\mathbf{u}} + \mathbf{u}'$ in a barotropic fluid, the sum of "mean" (overbar) and "fluctuation" (prime) velocity variances must be constant in time:

$$\frac{d}{dt} \iint \left(\overline{\mathbf{u}}^2 + (\mathbf{u}')^2 \right) d\mathbf{x} = 0,$$

for any perturbation field that is spatially orthogonal to the mean flow,

$$\iint \bar{\mathbf{u}} \cdot \mathbf{u}'\, d\mathbf{x} = 0.$$

(The orthogonality condition is satisfied for all the normal mode instabilities discussed in this chapter.) This implies that the kinetic energy associated with the fluctuations can grow only at the expense of the energy associated with the mean flow in the absence of any other flow components and that energy must be exchanged between these two components for this to occur. That is, there is a dynamical interaction between the mean flow and the fluctuations (also called eddies) that can be analyzed more generally than just for linear normal-mode fluctuations.

Again consider the particular situation of a parallel zonal flow (as in Section 3.3.3) with

$$\mathbf{u} = U(y, t)\,\hat{\mathbf{x}}. \tag{3.96}$$

In the absence of fluctuations or forcing, this is a stationary state (Section 3.1.4). For a small Rossby number, U is geostrophically balanced with a geopotential function,

$$\Phi(y, t) = -\int^{y} f(y')U(y', t)dy'.$$

Now, more generally, assume that there are fluctuations (designated by primes) around this background flow,

$$\mathbf{u} = \langle u \rangle(y, t)\,\hat{\mathbf{x}} + \mathbf{u}'(x, y, t), \qquad \phi = \langle \phi \rangle(y, t) + \phi'(x, y, t). \tag{3.97}$$

Here the angled bracket is defined as a zonal average. $\langle u \rangle$ is identified with U and $\langle \phi \rangle$ with Φ. With this definition for $\langle \cdot \rangle$, the average of a fluctuation field is zero, $\langle \mathbf{u}' \rangle = 0$; therefore, the KE orthogonality condition is satisfied. By substituting (3.97) into the barotropic equations and taking their zonal averages, the governing equations for $(\langle u \rangle, \langle \phi \rangle)$ are obtained. The mean continuity relation is satisfied exactly since $\partial_x \langle u \rangle$ is zero and $\langle v \rangle = 0$. The mean momentum equations are

$$\frac{\partial}{\partial t}\langle u \rangle = -\frac{\partial}{\partial y}\langle u'v' \rangle + \langle F^x \rangle$$

$$\frac{\partial \langle \phi \rangle}{\partial y} = -f\langle u \rangle - \frac{\partial}{\partial y}\langle v'^2 \rangle \tag{3.98}$$

after integrations by parts and subsitutions of the 2D continuity equation in (3.1). The possibility of a zonal-mean force, $\langle F^x \rangle$, is retained here, but $\langle F^y \rangle = 0$ is assumed, consistent with a forced zonal flow. All other terms from (3.1) vanish by the structure of the mean flow or by an assumption that the fluctuations are periodic, homogeneous (i.e., statistically invariant), or decaying away to zero in the zonal direction. The quadratic quantities, $\langle u'v' \rangle$ and $\langle v'^2 \rangle$, are zonally averaged eddy momentum fluxes due to products of fluctuation velocity.

The zonal mean flow is generally no longer a stationary state in the presence of the fluctuations. The first relation in (3.98) shows how the divergence of an eddy momentum flux, often called a *Reynolds stress*, can alter the mean flow or allow it to come to a new steady state by balancing its mean forcing. The second relation is a diagnostic one for the departure of $\langle \phi \rangle$ from its mean geostrophic component, again due to a Reynolds stress divergence. In the former relation, the indicated Reynolds stress, $\mathcal{R} = \langle u'v' \rangle$, is the mean meridional flux of zonal momentum by the fluctuations (eddies). In the latter relation, $\langle v'^2 \rangle = \langle v'v' \rangle$ is the mean meridional flux of meridional momentum by eddies.

As above, the kinetic energy can be written as the sum of mean and eddy energies,

$$\mathrm{KE} = \langle \mathrm{KE} \rangle + \mathrm{KE}' = \frac{1}{2} \int dy \left(\langle u \rangle^2 + \langle \mathbf{u}'^2 \rangle \right), \qquad (3.99)$$

since the cross term $\langle u \rangle u'$ vanishes by taking the zonal integral or average. The equation for $\langle \mathrm{KE} \rangle$ is derived by multiplying the zonal mean equation by $\langle u \rangle$ and integrating in y:

$$\frac{d}{dt} \langle \mathrm{KE} \rangle = - \int dy \langle u \rangle \frac{\partial}{\partial y} \langle u'v' \rangle + \int dy \langle u \rangle \langle F^x \rangle$$

$$= \int dy \langle u'v' \rangle \frac{\partial \langle u \rangle}{\partial y} + \int dy \langle u \rangle \langle F^x \rangle, \qquad (3.100)$$

assuming that $\langle u'v' \rangle$ and/or $\langle u \rangle$ vanish at the y boundaries. An analogous derivation for the eddy energy equation yields a compensating exchange (or *energy conversion*) term,

$$\frac{d}{dt} \mathrm{KE}' = - \int dy \langle u'v' \rangle \frac{\partial \langle u \rangle}{\partial y} + \int dy \langle \mathbf{u}' \cdot \mathbf{F}' \rangle, \qquad (3.101)$$

along with another term related to the fluctuating non-conservative force, $\mathbf{F}'$. Thus, the necessary and sufficient condition for KE' to grow at the expense of $\langle \mathrm{KE} \rangle$ is that the Reynolds stress, $\langle u'v' \rangle$, be anti-correlated on average (i.e., in a meridional integral) with the mean shear, $\partial_y \langle u \rangle$. This situation is often referred to as a *down-gradient eddy flux*. It is the most common paradigm for how eddies and mean flows influence each other: mean forcing generates mean flows that are then weakened or equilibrated by instabilities that generate eddies. If the forcing conditions are steady in time and some kind of statistical equilibrium is achieved for the flow as a whole, the eddies somehow achieve their own energetic balance between their generation by instability and a turbulent cascade to viscous *dissipation* (cf., Section 3.7). For example, if $\mathbf{F}'$ represents molecular viscous diffusion,

$$\mathbf{F}' = \nu\nabla^2\mathbf{u}',$$

then an integration by parts in (3.101) gives an integral relation,

$$\int dy\langle\mathbf{u}'\cdot\mathbf{F}'\rangle = -\int dy\,\nu\langle(\nabla\mathbf{u}')^2\rangle \leq 0,$$

that is never positive in the energy balance. The right-hand side here is called the energy dissipation. It implies a loss of KE′ whenever $\nabla\mathbf{u}' \neq 0$, and it at least has the correct sign to balance the energy conversion from the mean flow instability in the KE′ budget (3.101).

It can be shown that the barotropic instabilities in Section 3.3 all have down-gradient eddy momentum fluxes associated with the growing normal modes. (Note that the implied change in $\langle u\rangle$ from (3.98) is of the order of the fluctuation amplitude squared, $\mathcal{O}(\epsilon^2)$. Thus, for a normal-mode instability analysis, it is consistent to neglect any evolutionary change in $\langle u\rangle$ in the linearized equations for ψ' at $\mathcal{O}(\epsilon)$ when $\epsilon \ll 1$.)

3.5 Eddy viscosity and diffusion

The relation between the mean jet profile, $\langle u\rangle(y)$ and the Reynolds stress, $\langle u'v'\rangle(y)$ is illustrated in Fig. 3.13. The eddy flux is indeed directed opposite to the mean shear (i.e., it is down-gradient). Since a down-gradient eddy flux has the same sign as a mean viscous diffusion, $-\nu\partial_y\langle u\rangle$, these eddy fluxes can be anticipated to act in a way similar to viscosity, i.e., to smooth, broaden, and weaken the mean velocity profile, consistent with depleting $\langle KE\rangle$ and, in turn, generating KE′. This is expressed in a formula as

$$\langle u'v'\rangle \approx -\nu_e\frac{\partial}{\partial y}\langle u\rangle, \tag{3.102}$$

where $\nu_e > 0$ is the eddy viscosity coefficient. Equation (3.102) can either be viewed as a definition of $\nu_e(y)$ as a diagnostic measure of the eddy–mean inter-action or be utilized as a parameterization of the process with some specification of ν_e (Section 6.1.3). When this characterization is apt, the eddy–mean flow interaction is called an *eddy diffusion* process (also discussed in Chapters 5 and 6). Since in the present context the interaction occurs in the mean horizontal momentum balance, the process may more specifically be called horizontal *eddy viscosity* by analogy with molecular viscosity (Section 2.1.2), and the associated eddy viscosity coefficient is much larger than the molecular diffusivity, $\nu_e \gg \nu$,

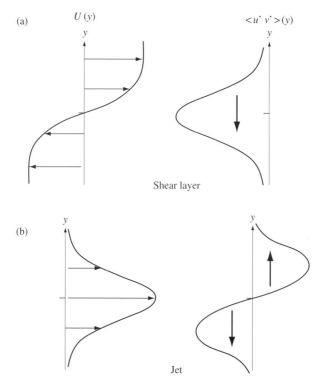

Fig. 3.13. Sketches of the mean zonal flow, $U(y)$ (left), and Reynolds stress profile, $\langle u'v'\rangle(y)$ (right), for (a) the mixing layer and (b) the Bickley Jet. Thin arrows on the left panels indicate the mean flow, and thick arrows on the right panels indicate the meridional flux of eastward zonal momentum. Thus, the eddy flux acts to broaden both the mixing layer and jet.

if advection by the velocity fluctuations acts much more rapidly to transport mean momentum than does the molecular viscous diffusion.

Eddy diffusion can be modeled for material tracers by analogy with a *random walk* for parcel trajectories and parcel tracer conservation. A random walk as a consequence of random velocity fluctuations is a simple, but crude, characterization of turbulence. Suppose that there is a large-scale mean tracer distribution, $\bar{\tau}(\mathbf{x})$, and fluctuations associated with the fluid motion, τ'. Further suppose that Lagrangian parcel trajectories have a mean and fluctuating component,

$$\mathbf{r} = \bar{\mathbf{r}} + \mathbf{r}'(t),\qquad(3.103)$$

and further suppose that an instantaneous tracer value is the same as its mean value at its mean location,

$$\tau(\mathbf{r}) = \bar{\tau}(\bar{\mathbf{r}}).\qquad(3.104)$$

The left-hand side can be decomposed into mean and fluctuation components, and a Taylor series expansion can be made about the mean parcel location,

$$\tau(\mathbf{r}) = \overline{\tau}(\overline{\mathbf{r}}+\mathbf{r}') + \tau'(\overline{\mathbf{r}}+\mathbf{r}') \approx \overline{\tau}(\overline{\mathbf{r}}) + (\mathbf{r}' \cdot \nabla)\overline{\tau}(\overline{\mathbf{r}}) + \tau'(\overline{\mathbf{r}}) + \cdots.$$

Substituting this into (3.104) yields an expression for the tracer fluctuation in terms of the trajectory fluctuation and the mean tracer gradient,

$$\tau' \approx -(\mathbf{r}' \cdot \nabla)\overline{\tau}, \tag{3.105}$$

after using the fact that the average of a fluctuation is zero. Now write an evolution equation for the large-scale tracer field, averaging over the space and time scales of the fluctuations,

$$\frac{\partial \overline{\tau}}{\partial t} + \overline{\mathbf{u}} \cdot \overline{\tau} = -\nabla \cdot (\overline{\mathbf{u}'\tau'}). \tag{3.106}$$

The right-hand side is the divergence of the eddy tracer flux. Substituting from (3.105) and using the trajectory evolution equation for $\mathbf{r}'$,

$$\overline{\mathbf{u}'\tau'} \approx -\overline{\mathbf{u}'(\mathbf{r}' \cdot \nabla)\overline{\tau}}$$

$$= -\overline{\frac{d\mathbf{r}'}{dt}(\mathbf{r}' \cdot \nabla)\overline{\tau}}$$

$$= -\left(\frac{d}{dt}\overline{\mathbf{r}'\mathbf{r}'}\right) \cdot \nabla\overline{\tau}. \tag{3.107}$$

An isotropic, random-walk model for trajectories assumes that the different coordinate directions are statistically independent and that the variance of parcel displacements, i.e., the *parcel dispersion*, $\overline{(r^{x'})^2} = \overline{(r^{y'})^2}$, increases linearly with time as parcels wander away from their mean location. This implies that

$$\frac{d}{dt}\overline{r^{i'}r^{j'}} = \kappa_e\,\delta_{i,j}, \tag{3.108}$$

where $\delta_{i,j}$ is the Kroneker delta function ($= 1$ if $i = j$ and $= 0$ if $i \neq j$) and i, j are coordinate direction indices. Here κ_e is called the Lagrangian *parcel diffusivity*, sometimes also called the Taylor diffusivity (after G. I. Taylor), and it is a constant in space and time for a random walk. Combining (3.106)–(3.108) gives the final form for the mean tracer evolution equation,

$$\frac{\partial \tau}{\partial t} + \mathbf{u} \cdot \nabla\tau = \kappa_e \nabla^2 \tau, \tag{3.109}$$

where the overbar averaging symbols are now implicit. Thus, if the fluctuating velocity field on small scales is random, then the effect on large-scale tracers is an eddy diffusion process. This type of turbulence parameterization is widely used in GFD, especially in general circulation models.

3.6 Emergence of coherent vortices

When a flow is barotropically unstable, its linearly unstable eigenmodes can amplify until the small-amplitude assumption of linearized dynamics is no longer valid. The subsequent evolution is nonlinear due to momentum advection. It can be correctly called a barotropic form of turbulence, involving *cascade* – the systematic transfer of fluctuation variance, such as the kinetic energy, from one spatial scale to another – *dissipation* – the removal of variance after a cascade carries it to a small enough scale so that viscous diffusion is effective – *transport* – altering the distributions of large-scale fields through stirring, mixing, and other forms of material rearrangement by the turbulent currents – and *chaos* – sensitive dependence and limited predictability from uncertain initial conditions or forcing (Section 3.2.2). Nonlinear barotropic dynamics also often leads to the emergence of *coherent vortices*, whose mutually induced movements and other more disruptive interactions can manifest all the attributes of turbulence.

Since these complex behaviors are difficult to capture with analytic solutions of (3.1), vortex emergence and evolution are illustrated here with several experimental and computational examples.

First consider a vortex sheet or free shear layer (Sections 3.2.1 and 3.3.3) with a small but finite thickness (Fig. 3.14). It is Kelvin–Helmholtz, unstable and zonally periodic fluctuations amplify in place as standing waves. Once the amplitude is large enough, an advective process of *axisymmetrization* begins to occur around each significant vorticity extremum in the fluctuation circulations. The fluctuation circulations all have the same parity, since their vorticity extrema have to come from the single-signed vorticity distribution of the parent shear layer. An axisymmetrization process transforms the spatial pattern of the fluctuations from a wave-like eigenmode (cf., (3.88)) toward a circular vortex (cf., Section 3.1.4). Once the vortices emerge, they move around under each other's influences, similar to point vortices (Section 3.2.1). As a result, pairing instabilities begin to occur where the nearest neighboring, like-sign vortices co-rotate (cf., Fig. 3.7, top) and deform each other. They move together and become intertwined in each other's vorticity distribution. This is called *vortex merger*. The evolutionary outcome of successive mergers between pairs of vortices is a vortex population with fewer vortices that have larger sizes and circulations. Finally, because of the sensitive dependence of these advective processes, the vortex motions are chaotic, and their spatial distribution becomes irregular, even when there is considerable regularity in the initial unstable mode.

For an unstable jet flow (Section 3.3.3), a similar evolutionary sequence occurs. However, since this mean flow has vorticity of both signs (Fig. 3.11), the vortices emerge with both parities. An experiment for a turbulent wake flow in a thin soap film that approximately mimics barotropic fluid dynamics is shown in Fig. 3.15.

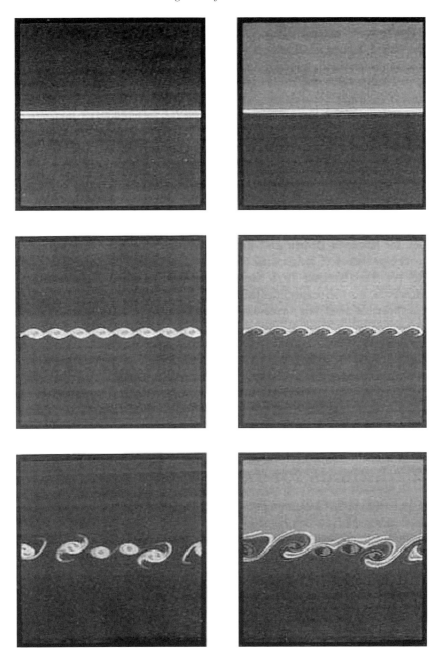

Fig. 3.14. Vortex emergence and evolution for a computational 2D parallel-flow shear layer with finite, but small, viscosity and tracer diffusivity. The two columns are for vorticity (left) and tracer (right), and the rows are successive times: near initialization (top); during the nearly linear, Kelvin–Helmholtz, varicose-mode, instability phase (middle); and after emergence of coherent anticyclonic vortices and approximately one cycle of pairing and merging of neighboring vortices (bottom). (Comte, 1989)

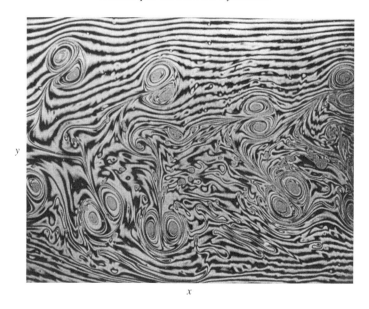

Fig. 3.15. Vortices after emergence and dipole pairing in an experimental 2D turbulent wake (i.e., jet). The stripes indicate approximate streamfunction contours. (Couder & Basdevant, 1986)

In this experiment a thin cylinder is dragged through the film, and this creates an unstable jet in its wake. The ensuing instability and vortex emergence leads to a population of vortices, many of which appear as *vortex couples* (i.e., dipole vortices that move as in Fig. 3.7, bottom).

3.7 Two-dimensional turbulence

Turbulence is an inherently dissipative phenomenon since advectively induced cascades spread the variance across different spatial scales, reaching down to arbitrarily small scales where molecular viscosity and diffusion can damp the fluctuations through mixing. Integral kinetic energy and *enstrophy* (i.e., vorticity variance) budgets can be derived from (3.1) with $\mathbf{F} = \nu\nabla^2\mathbf{u}$ and spatially periodic boundary conditions (for simplicity):

$$\frac{d\,\mathrm{KE}}{dt} = -\nu \iint dx\,dy (\boldsymbol{\nabla}\mathbf{u})^2$$

$$\frac{d\,\mathrm{Ens}}{dt} = -\nu \iint dx\,dy (\boldsymbol{\nabla}\zeta)^2. \tag{3.110}$$

KE is defined in (3.2), and

$$\mathrm{Ens} = \frac{1}{2} \iint dx\,dy\,\zeta^2. \tag{3.111}$$

Therefore, due to the viscosity, KE and Ens are non-negative quantities that are non-increasing with time as long as there is no external forcing of the flow.

The common means of representing the scale distribution of a field is through its *Fourier transform* and *spectrum*. For example, the Fourier integral for $\psi(\mathbf{x})$ is

$$\psi(\mathbf{x}) = \int d\mathbf{k}\, \hat{\psi}(\mathbf{k}) e^{i\mathbf{k}\cdot\mathbf{x}}. \tag{3.112}$$

$\mathbf{k}$ is the vector wavenumber, and $\hat{\psi}(\mathbf{k})$ is the complex Fourier transform coefficient. With this definition the spectrum of ψ is

$$S(\mathbf{k}) = \text{AVG}\left[|\hat{\psi}(\mathbf{k})|^2\right]. \tag{3.113}$$

The averaging is over any appropriate symmetries for the physical situation of interest (e.g., over time in a statistically stationary situation, over the directional orientation of $\mathbf{k}$ in an isotropic situation, or over independent realizations in a recurrent situation). $S(k)$ can be interpreted as the variance of ψ associated with a spatial scale, $L = 1/k$, with $k = |\mathbf{k}|$, such that the total variance, $\int d\mathbf{x}\,\psi^2$, is equal to $\int d\mathbf{k}\, S$ (sometimes called *Parceval's Theorem*).

With a Fourier representation, the energy and enstrophy are integrals over their corresponding spectra,

$$\text{KE} = \int d\mathbf{k}\,\text{KE}(\mathbf{k}), \quad \text{Ens} = \int d\mathbf{k}\,\text{Ens}(\mathbf{k}), \tag{3.114}$$

with

$$\text{KE}(\mathbf{k}) = \frac{1}{2}k^2 S, \quad \text{Ens}(\mathbf{k}) = \frac{1}{2}k^4 S = k^2\,\text{KE}(\mathbf{k}). \tag{3.115}$$

The latter relations are a consequence of the spatial gradient of ψ having a Fourier transform equal to the product of $\mathbf{k}$ and $\hat{\psi}$. The spectra in (3.115) have different shapes due to their different weighting factors of k, and the enstrophy spectrum has a relatively larger magnitude at smaller scales than does the energy spectrum (Fig. 3.16, top).

In the absence of viscosity – or during the early time interval after initialization with smooth, large-scale fields before the cascade carries enough variance to small scales to make the right-hand side terms in (3.110) significant – both KE and Ens are conserved with time. If the cascade process broadens the spectra (which is a generic behavior in turbulence, transferring variance among different spatial scales), the only way that both integral quantities can be conserved, given their different k weights, is that more of the energy is transferred toward larger scales (smaller k), while more of the enstrophy is transferred toward smaller scales (larger k). This behavior is firmly established by computational and laboratory studies, and it can at least partly be derived as a necessary consequence of spectrum broadening by the cascades. Define a *centroid wavenumber*, k_E (i.e., a

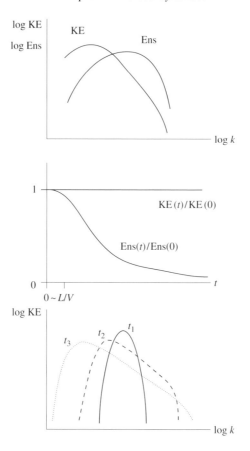

Fig. 3.16. (Top) schematic isotropic spectra for energy, KE(k), and enstrophy, Ens(k), in 2D turbulence at large Reynolds number. Note that the energy peak occurs at smaller k than the enstrophy peak. (Middle) time evolution of total energy, KE(t), and enstrophy, Ens(t), each normalized by their initial value. The energy is approximately conserved when $Re \gg 1$, but the enstrophy has significant decay over many eddy advective times, L/V. (Bottom) evolution of the energy spectrum, KE(k, t), at three successive times, $t_1 < t_2 < t_3$. With time the spectrum spreads, and the peak moves to smaller k.

characteristic wavenumber averaged across the spectrum), and a *wavenumber bandwidth*, Δk_E for the energy spectrum as follows:

$$k_E = \int d\mathbf{k} |\mathbf{k}| \mathrm{KE}(\mathbf{k}) / \mathrm{KE}$$

$$\Delta k_E = \left(\int d\mathbf{k} (|\mathbf{k}| - k_E)^2 \mathrm{KE}(\mathbf{k}) \right)^{1/2} / \mathrm{KE}. \qquad (3.116)$$

Both quantities are positive by construction. If the turbulent evolution broadens the spectrum, then conservation of KE and Ens (i.e., $\dot{\mathrm{KE}} = \dot{\mathrm{Ens}} = 0$, with the overlying dot again denoting a time derivative) implies that the energy centroid wavenumber must decrease,

$$\dot{\Delta k_E} > 0 \;\Rightarrow\; -2k_E\dot{k}_E > 0 \;\Rightarrow\; \dot{k}_E < 0.$$

This implies a systematic transfer of the energy toward larger scales. This tendency is accompanied by an increasing enstrophy centroid wavenumber, $\dot{k}_{Ens} > 0$ (with k_{Ens} defined analogously to k_E). These two, co-existing tendencies are referred to, respectively, as the *inverse energy cascade* and the *forward enstrophy cascade* of 2D turbulence. The indicated direction in the latter case is "forward" to small scales since this is the most common behavior in different regimes of turbulence (e.g., in 3D, uniform-density turbulence, Ens is not an inviscid integral invariant, and the energy cascade is in the forward direction).

In the presence of viscosity – or after the forward enstrophy cascade acts for long enough to make the dissipation terms become significant – KE will be much less efficiently dissipated than Ens because so much less of its variance – and the variance of the integrand in its dissipative term in the right-hand side of (3.110) – resides in the small scales. Thus, for large Reynolds number (small ν), Ens will decay significantly with time, while KE may not decay much at all (Fig. 3.16, middle). Over the course of time, the energy spectrum shifts toward smaller wavenumbers and larger scales, due to the inverse cascade, and its dissipation rate further declines (Fig. 3.16, bottom).

The cascade and dissipation in 2D turbulence co-exist with vortex emergence, movement, and mergers (Fig. 3.17). From smooth initial conditions, coherent vortices emerge by axisymmetrization, move approximately the same way as point vortices (Section 3.2), occasionally couple for brief intervals (Fig. 3.15), and merge when two vortices of the same parity move close enough together (Fig. 3.18). With time the vortices become fewer, larger, and sparser in space, and they undergo less frequent close encounters. Since close encounters are the occasions when the vortices change through deformation in ways other than simple movement, the overall evolutionary rates for the spectrum shape and vortex population become ever slower, even though the kinetic energy does not diminish. Enstrophy dissipation occurs primarily during emergence and merger events, as filaments of vorticity are stripped off the vortices. The filamentation is a consequence of the differential velocity field (i.e., shear, strain rate; Section 2.1.5), due to one vortex acting on another, that increases rapidly as the vortex separation distance diminishes. The filaments continue irreversibly to elongate until their transverse scale shrinks enough to come under the control of viscous diffusion, and the enstrophy they contain is thereby dissipated. So, the integral statistical outcomes of cascade and dissipation in 2D turbulence are the result of a sequence of local dynamical processes of the elemental coherent vortices, at least during

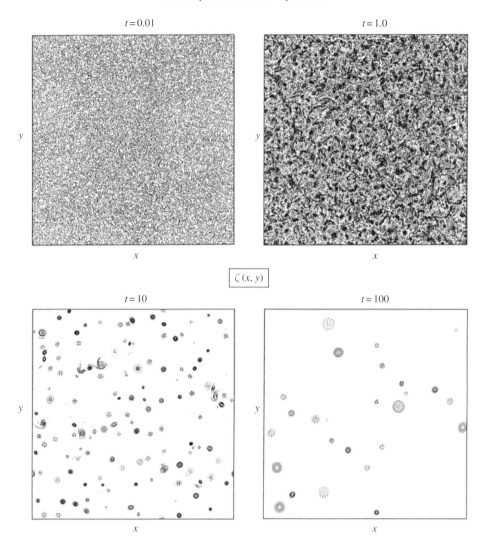

Fig. 3.17. Vortex emergence and evolution in computational 2D turbulence, as seen in $\zeta(x, y)$ at sequential times, with random, spatially smooth initial conditions. Solid contours are for positive ζ, and dashed ones are for negative ζ. The contour interval is twice as large in the first panel as in the others. The times are non-dimensional based on an advective scaling, L/V. (Adapted from McWilliams, 1984.)

the period after their emergence from complex initial conditions or forcing. The vortices substantially control the dynamics of 2D turbulent evolution.

The preceding discussion refers to freely evolving, or *decaying*, turbulence, where the ultimate outcome is energy and enstrophy dissipation through the inexorable action of viscosity. Alternatively, one can consider forced, *equilibrium*

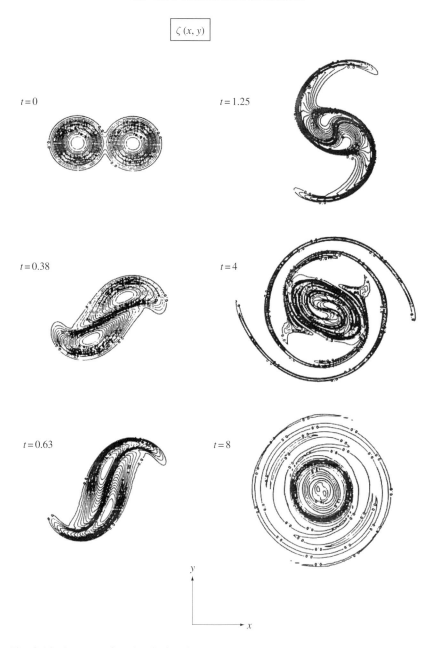

Fig. 3.18. Computational solution for the merger of two like-sign, bare-monopole vortices (in non-dimensional time units scaled by L/V) initially located near each other. The exterior strain field from each vortex deforms the vorticity distribution of the other one so that the ζ fields wrap around each other; their centers move together along spiral trajectories; and ultimately they blend together after viscous diffusion smooths the strong gradients. While this is occurring, vorticity filaments are cast off from the merging vortices, stretched by the exterior strain field, and dissipated by viscosity. (Adapted from McWilliams, 1991.)

turbulence, where a statistically stationary state occurs with the turbulent generation rate balancing the dissipation rate for energy, enstrophy, and all other statistical measures. The generation process may either be the instability of a forced mean flow (e.g., Section 3.4) or it may be an imposed fluctuating force. For 2D turbulence with fluctuations generated on intermediate spatial scales (smaller than the domain size and larger than the viscous scale), inverse energy and forward enstrophy cascades ensue. To achieve an equilibrium energy balance some dissipative process, beyond viscosity, must be included to deplete the energy at large scales, or else the energy will continue to grow. (A common choice is a linear drag force, motivated by the effect of an Ekman boundary layer; Chapters 5 and 6.) The degree of coherent vortex emergence and subsequent dynamical control of the equilibrium turbulence depends upon the relative rates of forcing and energy dissipation (which disrupt the vortex dynamics) and of vortex advection (which sustains it).

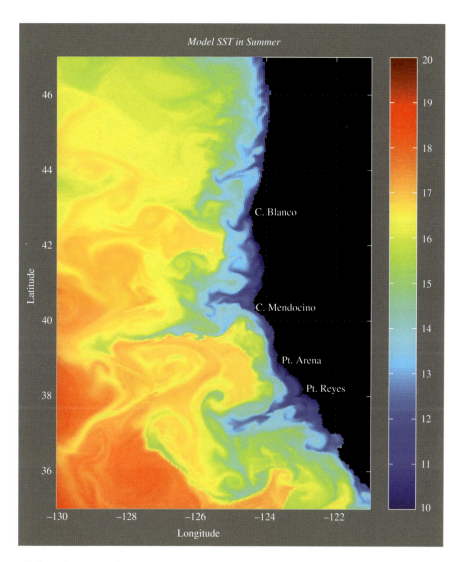

Color plate 1.   Fig. 1.2. Sea surface temperature off the US West Coast in late
summer, from a numerical oceanic model. Note the general pattern similarity with
Fig. 1.1 for cold upwelled water near the coastline, mesoscale vortices, and cold
filaments advected away from the boundary. However, the measured and simulated patterns should not be expected to agree in their individual features because
of the sensitive dependence of advective dynamics. (Marchesiello *et al.*, 2003.)

Color plate 2.   Fig. 4.1. Clouds in the stably stratified upper troposphere showing vapor condensation lines where an internal gravity wave has lifted air parcels.
The wave crests and troughs are aligned vertically in this photograph from the
ground, and the wave propagation direction is sideways.

Color plate 3.   Fig. 4.2. Oceanic internal gravity waves on the near-surface
pycnocline, as measured by a satellite's Synthetic Aperture Radar reflection from
the associated disturbances of the sea surface. The waves are generated by tidal
flow through the Strait of Gibraltar. (NASA Shuttle mission 41 G, 12:12:48,
Oct. 10, 1984)

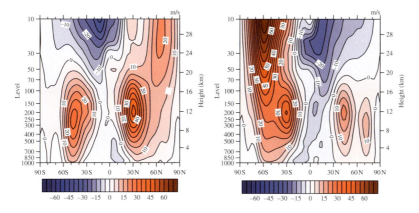

Color plate 4.   Fig. 5.1. The mean zonal wind [m s^{-1}] for the atmosphere, averaged over time and longitude during 2003: (left) January and (right) July. The
vertical axis is labeled both by height [km] and by pressure level [1 mb = 10^2 Pa].
Note the following features: the eastward velocity maxima at the tropopause (at
$\approx 200 \times 10^2$ Pa) that are stronger in the winter hemisphere; the wintertime stratospheric polar night jet; the stratospheric tropical easterlies that shift hemisphere
with the seasons; and the weak westward surface winds both in the tropics (i.e.,
trade winds) and near the poles. (National Centers for Environmental Prediction
climatological analysis (Kalnay *et al.*, 1996), courtesy of Dennis Shea, National
Center for Atmospheric Research.)

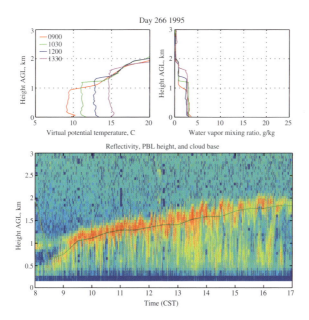

Color plate 5.   Fig. 6.1. Example of reflectivities (bottom) observed in the
cloud-free convective boundary layer in central Illinois on 23 September 1995:
(top left) virtual temperature profiles and (top right) vertical profiles of water
vapor mixing ratio. Note the progressive deepening of the layer through the
middle of the day as the ground warms (Angevine *et al.*, 1998).

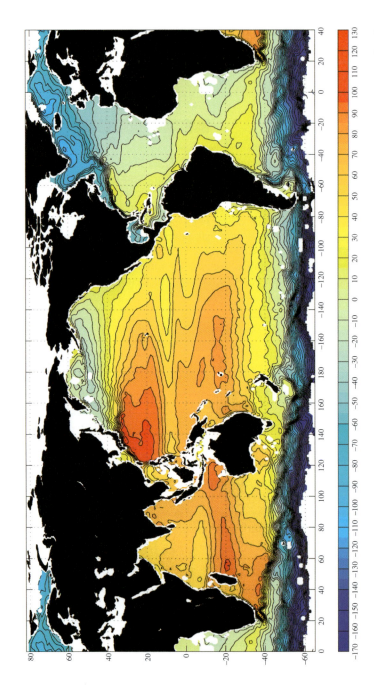

Color plate 6. Fig. 6.11. Observational estimate of time-mean sea level relative to a geopotential iso-surface, $\overline{\eta}$ [cm]. The estimate is based on near-surface drifting buoy trajectories, satellite altimetric heights, and climatological winds. $g\overline{\eta}/f$ can be interpreted approximately as the surface geostrophic streamfunction. Note the sub-tropical and sub-polar wind gyres with sea-level extrema adjacent to the continental boundaries on western sides of the major basins and the large sea-level gradient across the Antarctic Circumpolar Current (Niiler et al., 2003).

4

Rotating shallow-water and wave dynamics

Many types of wave motions occur in the ocean and atmosphere. They are characterized by an oscillatory structure in both space and time, as well as by systematic propagation from one place to another. Examples of the distinctive trough and crest patterns of internal gravity waves in the atmosphere and ocean are shown in Figs. 4.1 and 4.2.

In Chapters 2 and 3 some simple examples were presented for acoustic, internal-gravity, inertial, and Rossby wave oscillations. In this chapter a more extensive

Fig. 4.1. Color plate 2.   Clouds in the stably stratified upper troposphere showing vapor condensation lines where an internal gravity wave has lifted air parcels. The wave crests and troughs are aligned vertically in this photograph from the ground, and the wave propagation direction is sideways.

Fig. 4.2. Color plate 3. Oceanic internal gravity waves on the near-surface pycnocline, as measured by a satellite's Synthetic Aperture Radar reflection from the associated disturbances of the sea surface. The waves are generated by tidal flow through the Strait of Gibraltar. (NASA Shuttle mission 41G, 12:12:48, Oct. 10, 1984.)

examination is made for the latter three wave types and some others. This is done using a dynamical system that is more general than 2D fluid dynamics, because it includes a non-trivial influence of stable buoyancy stratification, but it is less general than 3D fluid dynamics. The system is called the *shallow-water equations*. In a strict sense, the shallow-water equations represent the flow in a fluid layer with uniform density, ρ_0, when the horizontal velocity is constant with depth (Fig. 4.3). This is most plausible for flow structures, where the horizontal scale, L, is much greater than the mean layer depth, H, i.e., $H/L \ll 1$. Recall from Section 2.3.4 that this relation is the same assumption that justifies the hydrostatic balance approximation, which is one of the ingredients in deriving the shallow-water equations. It is also correct to say that the shallow-water equations are a form of the hydrostatic primitive equations (Section 2.3.5) limited to a single degree of freedom in the vertical flow structure.

The shallow-water equations can therefore be interpreted literally as a model for barotropic motions in the ocean including the effects of its free surface. It is also representative of barotropic motions in the atmospheric troposphere, although

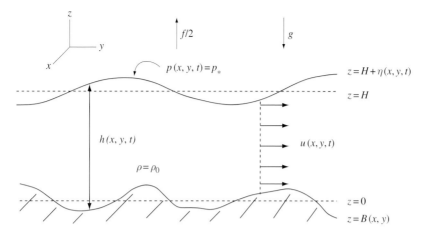

Fig. 4.3. Configuration for the shallow-water equations. They are valid for a fluid layer of uniform density, ρ_0, with an upper free surface where the pressure is p_*. The layer has a thickness, $h = H + \eta - B$; a depth-independent horizontal velocity, $\mathbf{u}$; a free surface elevation anomaly, η; and a bottom elevation, B. The mean positions of the top and bottom are $z = H$ and 0, respectively.

less obviously so because its upper free surface, the tropopause, may more readily influence and, in response, be influenced by the flows above it whose density is closer to the that of the troposphere than is true for air above water. The shallow-water equations mimic baroclinic motions, in a restricted sense, explained below, with only a single degree of freedom in their vertical structure (hence they are not fully baroclinic because $\hat{z} \cdot \nabla p \times \nabla \rho = 0$; Section 3.1.1). Nevertheless, in GFD there is a long history of accepting the shallow-water equations as a relevant analog dynamical system for some baroclinic processes. This view rests on the experience that shallow-water equations solutions have useful qualitative similarities with some solutions for 3D stably stratified fluid dynamics in, say, the Boussinesq or primitive equations. The obvious advantage of the shallow-water equations, compared to 3D equations, is their 2D spatial dependence, hence their greater mathematical and computational simplicity.

4.1 Rotating shallow-water equations

The fluid layer thickness is expressed in terms of the mean layer depth, H, upper free surface displacement, $\eta(x, y, t)$, and topographic elevation of the solid bottom surface, $B(x, y)$:

$$h = H + \eta - B. \tag{4.1}$$

Obviously, $h > 0$ is a necessary condition for shallow-water equations to have a meaningful solution. The kinematic boundary conditions (Section 2.1.1) at the top and bottom surfaces of the layer are

$$w = \frac{D(H + \eta)}{Dt} = \frac{D\eta}{Dt} \quad \text{at} \quad z = H + \eta$$

$$w = \frac{DB}{Dt} = \mathbf{u} \cdot \boldsymbol{\nabla} B \quad \text{at} \quad z = B, \tag{4.2}$$

respectively, where the vector quantities are purely horizontal. Since $\partial_z(u, v) = 0$ by assumption, the incompressible continuity relation implies that w is a linear function of z. Fitting this form to (4.2) yields

$$w = \left(\frac{z - B}{h}\right)\frac{D\eta}{Dt} + \left(\frac{h + B - z}{h}\right)\mathbf{u} \cdot \boldsymbol{\nabla} B. \tag{4.3}$$

Consequently,

$$\frac{\partial w}{\partial z} = \frac{1}{h}\frac{D\eta}{Dt} - \frac{1}{h}\mathbf{u} \cdot \boldsymbol{\nabla} B$$

$$= \frac{1}{h}\frac{D(\eta - B)}{Dt}$$

$$= \frac{1}{h}\frac{Dh}{Dt}. \tag{4.4}$$

Combining this with the continuity equation gives

$$\frac{\partial w}{\partial z} = -\boldsymbol{\nabla} \cdot \mathbf{u} = \frac{1}{h}\frac{Dh}{Dt}$$

$$\implies \frac{Dh}{Dt} + h\boldsymbol{\nabla} \cdot \mathbf{u} = 0$$

$$\text{or} \quad \frac{\partial h}{\partial t} + \boldsymbol{\nabla} \cdot (h\mathbf{u}) = 0. \tag{4.5}$$

This is called the *height* or *thickness equation* for h in the shallow-water equations. It is a vertically integrated expression of local mass conservation: the surface elevation goes up and down in response to the depth-integrated convergence and divergence of fluid motions (cf., integral mass conservation; Section 4.1.1).

The free-surface boundary condition on pressure (Section 2.2.3) is $p = p_*$, a constant; this is equivalent to saying that any fluid motion above the layer under consideration is negligible in its conservative dynamical effects on this layer (note, a possible non-conservative effect, also being neglected here, is a surface viscous stress). Integrate the hydrostatic relation downward from the surface, assuming uniform density, to obtain the following:

$$\frac{\partial p}{\partial z} = -g\rho = -g\rho_0 \tag{4.6}$$

$$\Longrightarrow p(x, y, z, t) = p_* + \int_z^{H+\eta} g\rho_0 \, dz'$$

$$\Longrightarrow p = p_* + g\rho_0(H + \eta - z). \tag{4.7}$$

In the horizontal momentum equations the only aspect of p that matters is its horizontal gradient. From (4.7),

$$\frac{1}{\rho_0}\mathbf{\nabla}p = g\mathbf{\nabla}\eta;$$

hence,

$$\frac{D\mathbf{u}}{Dt} + f\hat{\mathbf{z}} \times \mathbf{u} = -g\mathbf{\nabla}\eta + \mathbf{F}. \tag{4.8}$$

The equations (4.1), (4.5), and (4.8) comprise the shallow-water equations and are a closed partial differential equation system for $\mathbf{u}$, h, and η.

An alternative conceptual basis for the shallow-water equations is the configuration sketched in Fig. 4.4. It is for a fluid layer beneath a flat, solid, top boundary and with a deformable lower boundary separating the active fluid layer above from an inert layer below. For example, this is an idealization of the oceanic *pycnocline* (often called the *thermocline*), a region of strongly stable density stratification beneath the weakly stratified upper ocean region, which contains, in particular, the often well mixed surface boundary layer (cf., Chapter 6), and above the thick, weakly stratified abyssal ocean (Fig. 2.7). Accompanying approximations in this conception are a rigid lid (Section 2.2.3) and negligibly weak abyssal flow at greater depths. Again integrate the hydrostatic relation down from the upper surface, where $p = p_u(x, y, t)$ at $z = 0$, through the active layer, across its lower interface at $z = -(H + b)$ into the inert lower layer, to obtain the following:

$$p = p_u - g\rho_0 z, \quad -(H+b) \leq z \leq 0$$

$$p = p_l = g\rho_0(H+b) + p_u, \quad z = -(H+b)$$

$$p = p_l - g\rho_1(H+b+z), \quad z \leq -(H+b) \tag{4.9}$$

(using the symbols defined in Fig. 4.4). For the lower layer (i.e., $z \leq -(H+b)$) to be inert, $\mathbf{\nabla}p$ must be zero for a consistent force balance there. Hence,

$$\mathbf{\nabla}p_i = g\rho_1 \mathbf{\nabla}b, \tag{4.10}$$

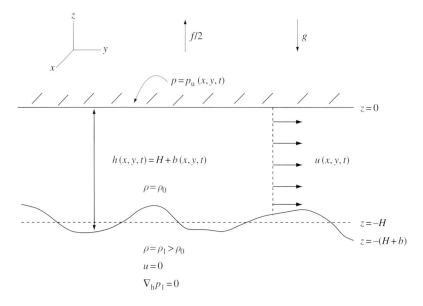

Fig. 4.4. Alternative configuration for the shallow-water equations with a rigid lid and a lower free interface above a motionless lower layer. ρ_0, $\mathbf{u}$, and h have the same meaning as in Fig. 4.3. Here p_{u} is the pressure at the lid; $-b$ is elevation anomaly of the interface; ρ_1 is the density of the lower layer; and $g' = g(\rho_1 - \rho_0)/\rho_0$ is the reduced gravity. The mean positions of the top and bottom are $z = 0$ and $z = -H$, respectively.

and

$$\nabla p_{\mathrm{u}} = g(\rho_1 - \rho_0)\nabla b = g'\rho_0 \nabla b. \qquad (4.11)$$

In (4.11),

$$g' = g\frac{\rho_1 - \rho_0}{\rho_0} \qquad (4.12)$$

is called the *reduced gravity* appropriate to this configuration, and the shallow-water equations are sometimes called the reduced-gravity equations.

The shallow-water equations corresponding to Fig. 4.4 are isomorphic to those for the configuration in Fig. 4.3 with the following identifications:

$$(b, g', 0) \longleftrightarrow (\eta, g, B), \qquad (4.13)$$

i.e., for the special case of the bottom being flat in Fig. 4.3. In the following, for specificity, the shallow-water equation notation used will be the same as in Fig. 4.3.

4.1.1 Integral and parcel invariants

Consider some of the conservative *integral invariants* for the shallow-water equations with $\mathbf{F} = 0$.

The total mass of the uniform-density, shallow-water fluid, $\rho_0 M$, is related to the layer thickness by

$$M = \iint dx\,dy\; h. \tag{4.14}$$

Mass conservation is derived by spatially integrating (4.5) and making use of the kinematic boundary condition (i.e., the normal velocity vanishes at the side boundary, denoted by $\mathcal{C}$):

$$\iint dx\,dy\frac{\partial h}{\partial t} = -\iint dx\,dy\; \mathbf{\nabla}\cdot (h\mathbf{u})$$

$$\implies \frac{dM}{dt} = \frac{d}{dt}\iint dx\,dy\; \eta$$

$$= -\oint_{\mathcal{C}} ds\;(h\mathbf{u})\cdot\hat{\mathbf{n}} = 0 \tag{4.15}$$

since both H and B are independent of time. H is defined as the average depth of the fluid over the domain,

$$H = \frac{1}{\text{area}}\iint dx\,dy\, h$$

so that η and B represent departures from the average heights of the surface and bottom.

Energy conservation is derived by the following operation on the momentum and thickness relations of the shallow-water equations, (4.8) and (4.5):

$$\iint dx\,dy\left(h\mathbf{u}\cdot(\textbf{momentum}) + \left[g\eta + \frac{1}{2}u^2\right](\textbf{thickness})\right). \tag{4.16}$$

With compatible boundary conditions that preclude advective fluxes through the side boundaries, this expression can be manipulated to derive

$$\frac{dE}{dt} = 0, \quad E = \iint dx\,dy\; \frac{1}{2}(h\mathbf{u}^2 + g\eta^2). \tag{4.17}$$

Here the total energy, E, is the sum of two terms, *kinetic energy* and *potential energy*. Only the combined energy is conserved, and exchange between the kinetic and potential components is freely allowed (and frequently occurs pointwise among the integrands in (4.17) for most shallow-water equations wave types).

The potential energy in (4.17) can be related to its more fundamental definition for a Boussinesq fluid (2.19),

$$\text{PE} = \frac{1}{\rho_o} \iiint dx\,dy\,dz\,\rho g z. \tag{4.18}$$

For a shallow-water fluid with constant $\rho = \rho_o$, the vertical integration can be performed explicitly to yield

$$\text{PE} = \iint dx\,dy\,\frac{1}{2}g z^2 \Big|_B^{H+\eta}$$

$$= \frac{g}{2} \iint dx\,dy\,[H^2 + 2H\eta + \eta^2 - B^2]. \tag{4.19}$$

Since both H and B are independent of time and $\iint dx\,dy\,\eta = 0$ by the defintion of H after (4.15),

$$\frac{d}{dt}\text{PE} = \frac{d}{dt}\text{APE}, \tag{4.20}$$

where

$$\text{APE} = \frac{1}{2}g \iint dx\,dy\,\eta^2 \tag{4.21}$$

is the same quantity that appears in (4.17). APE is called *available potential energy* since it is the only part of the PE that can change with time and thus is available for conservative dynamical exchanges with the KE. The difference between PE and APE is called *unavailable potential energy*, and it does not change with time for adiabatic dynamics. Since usually $H \gg |\eta|$, the unavailable part of the PE in (4.19) is much larger than the APE, and this magnitude discrepancy is potentially confusing in interpreting the energetics associated with the fluid motion (i.e., the KE). This concept can be generalized to 3D fluids, and it is the usual way that the energy balances of the atmospheric and oceanic general circulations are expressed.

There is another class of invariants associated with the *potential vorticity*, q (cf., Section 3.1.2). The dynamical equation for q is obtained by taking the curl of (4.8) (as in Section 3.1.2):

$$\frac{D\zeta}{Dt} + \mathbf{u} \cdot \nabla f + (f + \zeta)\nabla \cdot \mathbf{u} = \mathcal{F}, \tag{4.22}$$

or, by substituting for $\nabla \cdot \mathbf{u}$ from the second relation in (4.5),

$$\frac{D(f + \zeta)}{Dt} - \frac{f + \zeta}{h}\frac{Dh}{Dt} = \mathcal{F} \tag{4.23}$$

$$\implies \frac{Dq}{Dt} = \frac{1}{h}\mathcal{F}, \quad q = \frac{f(y) + \zeta}{h}. \tag{4.24}$$

Thus, q is again a *parcel invariant* for conservative dynamics, though it has a more general definition in the shallow-water equations than in the 2D definition (3.28).

In the shallow-water equations, in addition to the relative and planetary vorticity components present in 2D potential vorticity (ζ and $f(y)$, respectively), q now also contains the effects of *vortex stretching*. The latter can be understood in terms of the Lagrangian conservation of circulation, as in Kelvin's circulation theorem (Section 3.1.1). For a material parcel with the shape of an infinitesimal cylinder (Fig. 4.5), the local value of *absolute vorticity*, $f+\zeta$, changes with the thickness of the cylinder, h, while preserving the volume element of the cylinder, $h\,d$ area, so that the ratio of $f+\zeta$ and h (i.e., the potential vorticity, q) is conserved following the flow. For example, stretching the cylinder (h increasing and d area decreasing) causes an increase in the absolute vorticity ($f+\zeta$ increasing). This would occur for a parcel that moves over a bottom depression and thereby develops a more cyclonic circulation as long as its surface elevation, η, does not decrease as much as B does.

The conservative integral invariants for potential vorticity are derived by the following operation on (4.24) and (4.5):

$$\iint dx\,dy\ \left(nhq^{n-1}\cdot(\text{potential vorticity})+q^{n}\cdot(\text{thickness})\right)$$

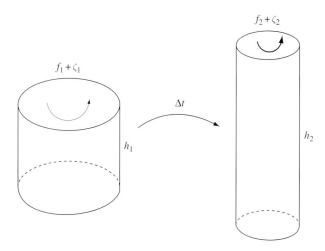

Fig. 4.5. Vortex stretching and potential vorticity conservation. If a material column is stretched to a greater thickness ($h_2 > h_1 > 0$) while conserving its volume, the potential vorticity conservation, $q_2 = q_1 > 0$ implies an increase in the absolute vorticity, $f(y_2)+\zeta_2 > f(y_1)+\zeta_1 > 0$.

for any value of n, or

$$\iint dx \, dy \left(nhq^{n-1} \left[\frac{\partial q}{\partial t} + \mathbf{u} \cdot \nabla q \right] + q^n \left[\frac{\partial h}{\partial t} + \nabla \cdot (h\mathbf{u}) \right] \right) = 0.$$

Since

$$\iint dx \, dy \, \nabla \cdot (A\mathbf{u}) = \int ds \, A\mathbf{u} \cdot \hat{\mathbf{n}} = 0,$$

for A an arbitrary scalar, if $\mathbf{u} \cdot \hat{\mathbf{n}} = 0$ on the boundary (i.e., the kinematic boundary condition of zero normal flow at a solid boundary), the result is

$$\frac{d}{dt} \iint dx \, dy \, hq^n = 0. \tag{4.25}$$

This result is identical to that of 2D flows (3.29), so again it is true that integral functionals of q are preserved under conservative evolution. This is because the fluid motion can only rearrange the locations of the parcels with their associated q values by (4.24), but it cannot change their q values. The same rearrangement principle and integral invariants are true for a passive scalar field (assuming it has a uniform vertical distribution for consistency with the shallow-water equations), ignoring any effects from horizontal diffusion or side-boundary flux. The particular invariant for $n = 2$ is called *potential enstrophy*, analogous to enstrophy as the integral of vorticity squared (Section 3.7).

4.2 Linear wave solutions

Now consider the *normal-mode wave solutions* for the shallow-water equations with $f = f_0$, $B = 0, \mathbf{F} = 0$, and an unbounded domain. These are solutions of the dynamical equations linearized about a state of rest with $\mathbf{u} = \eta = 0$, so they are appropriate dynamical approximations for small-amplitude flows. The linear shallow-water equations from (4.5) and (4.8) are

$$\frac{\partial u}{\partial t} - fv = -g \frac{\partial \eta}{\partial x}$$

$$\frac{\partial v}{\partial t} + fu = -g \frac{\partial \eta}{\partial y}$$

$$\frac{\partial \eta}{\partial t} + H \left(\frac{\partial u}{\partial x} + \frac{\partial v}{\partial y} \right) = 0. \tag{4.26}$$

These equations can be combined to leave η as the only dependent variable (or, alternatively, u or v): first form the combinations,

$$\partial_t (1\text{st}) + f(2\text{nd}) \longrightarrow (\partial_{tt} + f^2)u$$

$$= -g(\partial_{xt}\eta + f\partial_y \eta)$$

$$\partial_t(\text{2nd}) - f(\text{1st}) \longrightarrow (\partial_{tt} + f^2)v$$

$$= -g(\partial_{yt}\eta - f\partial_x\eta)$$

$$(\partial_{tt} + f^2)(\text{3rd}) \longrightarrow (\partial_{tt} + f^2)\partial_t\eta$$

$$= -H(\partial_{tt} + f^2)(\partial_x u + \partial_y v), \tag{4.27}$$

then substitute the x- and y-derivatives of the first two relations into the last relation,

$$\left[\frac{\partial^2}{\partial t^2} + f^2\right]\frac{\partial\eta}{\partial t} = gH\left(\frac{\partial^3\eta}{\partial x^2\partial t} + f\frac{\partial^2\eta}{\partial y\partial x} + \frac{\partial^3\eta}{\partial y^2\partial t} - f\frac{\partial^2\eta}{\partial y\partial x}\right)$$

$$\Longrightarrow \frac{\partial}{\partial t}\left[\frac{\partial^2}{\partial t^2} + f^2 - gH\nabla^2\right]\eta = 0. \tag{4.28}$$

This combination thus results in a partial differential equation for η alone. The normal modes for (4.26) or (4.28) have the form

$$[u, v, \eta] = \text{Real}\left([u_0, v_0, \eta_0]e^{i(\mathbf{k}\cdot\mathbf{x} - \omega t)}\right). \tag{4.29}$$

When (4.29) is inserted into (4.28), the partial differential equation becomes an algebraic equation:

$$-i\omega(-\omega^2 + f^2 + gH\mathbf{k}^2)\eta_0 = 0, \tag{4.30}$$

or for $\eta_0 \neq 0$, divide by $-i\eta_0$ to obtain

$$\omega(\omega^2 - [f^2 + c^2\mathbf{k}^2]) = 0. \tag{4.31}$$

The quantity

$$c = \sqrt{gH} \tag{4.32}$$

is a gravity wave speed (Sections 4.2.2 and 4.5). Equation (4.31) is called the *dispersion relation* for the linear shallow-water equations (cf., the dispersion relation for a Rossby wave; Section 3.1.2). It has the generic functional form for waves, $\omega = \omega(\mathbf{k})$. Here the dispersion relation is a cubic equation for the eigenvalue (or *eigenfrequency*) ω; hence there are three different wave eigenmodes for each $\mathbf{k}$.

Wave propagation

The dispersion relation determines the propagation behavior for waves. Any quantity with an exponential space-time dependence as in (4.29) is spatially uniform in the direction perpendicular to $\mathbf{k}$ at any instant, and its spatial pattern propagates parallel to $\mathbf{k}$ at the *phase velocity* defined by

$$\mathbf{c}_\mathrm{p} = \frac{\omega}{\mathbf{k}^2}\mathbf{k}. \tag{4.33}$$

However, the pattern shape is not necessarily preserved during an extended propagation interval (i.e., over many wavelengths, $\lambda = 2\pi/|\mathbf{k}|$, and/or many wave periods, $P = 2\pi/|\omega|$). If the spatial pattern is a superposition of many different component wavenumbers (e.g., as in a Fourier transform; Section 3.7), and if the different wavenumber components propagate at different speeds, then their resulting superposition will yield a temporally changing shape. This process of wavenumber separation by propagation is called *wave dispersion*. If the pattern has a dominant wavenumber component, $\mathbf{k}_*$, and its amplitude (i.e., the coefficient of the exponential function in (4.29)) is spatially localized within some region that is large compared to $\lambda_* = 2\pi/k_*$, then the region that has a significant wave amplitude will propagate with the *group velocity* defined by

$$\mathbf{c}_{\mathrm{g}} = \left.\frac{\partial\omega}{\partial\mathbf{k}}\right|_{\mathbf{k}=\mathbf{k}_*}. \tag{4.34}$$

Thus, one can say that the wave energy propagates with $\mathbf{c}_{\mathrm{g}}$, not $\mathbf{c}_{\mathrm{p}}$. If $\mathbf{c}_{\mathrm{p}} \neq \mathbf{c}_{\mathrm{g}}$, the pattern shape will evolve within this region through dispersion, but if these two wave velocities are equal then the pattern shape will be preserved with propagation. Waves with a dispersion relation that implies $\mathbf{c}_{\mathrm{p}} = \mathbf{c}_{\mathrm{g}}$ are called *non-dispersive*. There is extensive scientific literature on the many types of waves that occur in different media; e.g., Lighthill (1978) and Pedlosky (2003) are relevant books about waves in GFD.

4.2.1 Geostrophic mode

The first eigenvalue in (4.31) is

$$\omega = 0; \tag{4.35}$$

i.e., it has neither phase nor energy propagation. From (4.29) and (4.26), this mode satisfies the relations

$$fv_0 = +ikg\eta_0$$
$$fu_0 = -i\ell g\eta_0$$
$$iku_0 + i\ell v_0 = 0 \tag{4.36}$$

for $\mathbf{k} = (k, \ell)$. Note that this is geostrophic motion. It is horizontally non-divergent and has a streamfunction modal amplitude,

$$\psi_0 = g\frac{\eta_0}{f}$$

(cf., Section 2.4.2). In the linear, conservative shallow-water equations (4.26), the geostrophic mode is a stationary solution (with $\partial_t = 0$).

4.2.2 Inertia-gravity waves

The other two eigenfrequency solutions for (4.31) have $\omega \neq 0$:

$$\omega^2 = [f^2 + c^2 \mathbf{k}^2]$$

$$\implies \omega = \pm[f^2 + c^2 K^2]^{1/2}, \quad K = |\mathbf{k}|. \tag{4.37}$$

First take the *long-wave limit* ($\mathbf{k} \to 0$):

$$\omega \to \pm f. \tag{4.38}$$

These are *inertial waves* (cf., Section 2.4.3). The phase velocity, $\mathbf{c}_\mathrm{p} = \omega \mathbf{k}/K^2 \to f\mathbf{k}/K^2 \to \infty$. Thus, the phase propagation becomes infinitely fast in this inertial-wave limit.

Alternatively, take the *short-wave limit* ($\mathbf{k} \to \infty$):

$$\omega \to \pm cK \to \infty, \tag{4.39}$$

whose phase velocity, $\mathbf{c}_\mathrm{p} \to c\hat{\mathbf{e}}_\mathbf{k}$, remains finite with a speed c in the direction of the wavenumber vector, $\hat{\mathbf{e}}_\mathbf{k} = \mathbf{k}/K$. Waves in the limit (4.39) are non-dispersive. Since any initial condition can be represented as a superposition of $\mathbf{k}$ components by a Fourier transform (Section 3.7), it will preserve its shape during propagation. In contrast, waves near the inertial limit (4.38) are highly dispersive and do not preserve their shape.

For the linear shallow-water equations, the Brünt–Väisällä frequency (Section 2.3.3) is evaluated as

$$N^2 = -\frac{g}{\rho}\frac{\Delta\rho}{\Delta z} = -\frac{g}{\rho_0}\left(\frac{0 - \rho_0}{H}\right) = \frac{g}{H}. \tag{4.40}$$

Thus, for the short-wave limit,

$$|\omega| = cK = \sqrt{gH}\, K = NKH. \tag{4.41}$$

Recall that the shallow-water equations are a valid approximation to the more generally 3D motion in a uniform-density fluid layer only for $H/L \ll 1$, or equivalently $KH \ll 1$. Thus,

$$|\omega| \to N \quad KH \to 1^-$$

in (4.41); this, rather than $KH \to \infty$ and the resulting (4.39), is about as far as the short-wave limit should be taken for the shallow-water equations, due to the derivational assumption of hydrostatic balance and thinness, $H/L \ll 1$ (Section 4.1). Recall from Section 2.3.3 that $\omega = \pm N$ is the frequency for an internal gravity oscillation in a stably stratified 3D fluid. (In fact, this is the largest internal gravity wave frequency in a 3D Boussinesq equations normal-mode solution; NB, Exercise 6 for this chapter.) The limit (4.39) is identified as the *gravity-wave* mode for the shallow-water equations. It can be viewed alternatively as an *external* or a *surface gravity wave* for a water layer beneath a vacuum or an air layer (Fig. 4.3), or as an *internal gravity wave* on an interface with the appropriately reduced gravity, g', and buoyancy frequency, N (Fig. 4.4).

It is typically true that "deep" gravity waves with a relatively large vertical scale, comparable to the depth of the pycnocline or tropopause, have a faster phase speed, c, than the parcel velocity, V. Their ratio is called the *Froude number*,

$$Fr = \frac{V}{c} = \frac{V}{\sqrt{gH}} = \frac{V}{NH}.$$ (4.42)

Deep internal gravity wave speeds are typically $\mathcal{O}(10^2)$ m s^{-1} in the atmosphere and $\mathcal{O}(1)$ m s^{-1} in the ocean. For the V values characterizing large-scale flows (Section 2.4.2), the corresponding Froude numbers are $Fr \sim Ro$ in both media. Thus, these gravity waves are rapidly propagating in comparison to advective parcel movements, but also recall that sound waves are even faster than gravity waves, with $M \ll Fr$ (Section 2.2.2).

Based on the short- and long-wave limits (4.38) and (4.39), the second set of modes (4.37) are called *inertia-gravity waves*, or, in the terminology of Pedlosky (Section 3.9, 1987), *Poincaré waves*. Note that these modes are horizontally isotropic because their frequency and phase speed, $|\mathbf{c}_p|$, are independent of the propagation direction, $\hat{\mathbf{e}}_k$, since (4.37) depends only on the wavenumber magnitude, K, rather than $\mathbf{k}$ itself.

For inertia-gravity waves the approximate boundary between the predominantly inertial and gravity wave behaviors occurs for $KR = 1$, where

$$R = \frac{c}{|f|} = \frac{\sqrt{gH}}{f} = \frac{NH}{f}$$ (4.43)

is the *radius of deformation* (sometimes called the *Rossby radius*). R is commonly an important length scale in rotating, stably stratified fluid motions, and many other examples of its importance will be presented later. In the context of the rigid-lid approximation, R is the *external deformation radius*, R_e in (2.113), associated with the oceanic free surface. R in (4.43) has the same interpretation for the shallow-water equations configuration with full gravitational acceleration, g (Fig. 4.3), but it should alternatively be interpreted as an *internal deformation radius* with the reduced gravity, g', representing the interior stratification in the configuration in Fig. 4.4, as well as in 3D stratified fluids (Chapter 5). Internal deformation radii are much smaller than external ones because $g' \ll g$; typical values are several 100 km in the troposphere and several 10 km in the ocean.

For the inertia-gravity modes, the modal amplitude for vorticity is

$$\zeta_0 = ikv_0 - i\ell u_0$$

$$= -\frac{gfK^2}{\omega^2 - f^2}\eta_0 = \frac{gf}{c^2}\eta_0, \tag{4.44}$$

using the relations following (4.26), the modal form (4.29), and the dispersion relation (4.37). A linearized approximation of q from (4.24) is

$$q - \frac{f}{H} = \frac{f + \zeta}{H + \eta} - \frac{f}{H} \approx \frac{\zeta}{H} - \frac{f\eta}{H^2}. \tag{4.45}$$

Hence, the modal amplitude for inertia-gravity waves is

$$q_0 = \frac{\zeta_0}{H} - \frac{f\eta_0}{H^2}$$

$$= \frac{gf}{Hc^2}\eta_0 - \frac{f}{H^2}\eta_0 = 0, \tag{4.46}$$

using (4.44) for ζ_0. Thus, these modes have no influence on the potential vorticity, which is entirely carried by the geostrophic modes whose modal q amplitude is

$$q_0 = -\frac{g}{fH}[R^{-2} + K^2]\eta_0 \neq 0. \tag{4.47}$$

4.2.3 Kelvin waves

There is an additional type of wave mode for the linear shallow-water equations (4.26) when a side boundary is present. This is illustrated for a straight wall at $x = 0$ (Fig. 4.6), where the kinematic boundary condition is $u = 0$. The normal-mode solution and dispersion relation are

$$u = 0$$

$$v = -\frac{g}{fR}\eta_0 e^{-x/R}\sin[\ell y - \omega t]$$

$$\eta = \eta_0 e^{-x/R}\sin[\ell y - \omega t]$$

$$\omega = -fR\ell, \tag{4.48}$$

as can be verified by substitution into (4.26). This eigensolution is called a *Kelvin wave*. It is non-dispersive since ω/ℓ is a constant. It stays trapped against the boundary with the off-shore decay scale, R; it oscillates in time with frequency, ω; and it propagates along the boundary with the gravity-wave speed, c (since $fR = \text{sign}[f]\sqrt{gH}$) in the direction that has the boundary located to the right in

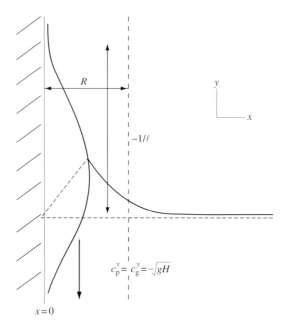

Fig. 4.6. A mid-latitude, f-plane Kelvin wave along a western boundary in the northern hemisphere. ℓ is the along-shore wavenumber, R is the deformation radius, and c_p^y and c_g^y are the meridional phase and group velocities. The wave propagates southward at the shallow-water gravity wave speed.

the northern hemisphere (i.e., it circles around the bounded domain in a cyclonic sense). The cross-shore momentum balance is geostrophic,

$$-fv = -g\frac{\partial \eta}{\partial x}; \tag{4.49}$$

whereas the along-shore momentum balance is the same as in a pure gravity wave,

$$v_t = -g\frac{\partial \eta}{\partial y}. \tag{4.50}$$

Thus, the dynamics of a Kelvin wave is a hybrid combination of the influences of rotation and stratification.

The ocean is full of Kelvin waves near the coasts, generated as part of the response to changing wind patterns (although their structure and propagation speed are usually modified from the solution (4.48) by the cross-shore bottom-topographic profile). A particular example of this occurs as a consequence of the evolution of El Niño (Fig. 4.7). Equatorial fluctuations near the eastern boundary generate poleward- (i.e., cyclonic-) propagating Kelvin waves along the eastern boundary within a layer whose width is the local deformation radius near the Equator,

$$R(y) = \frac{NH}{|f(y)|} \approx \frac{N_0 H_0 a}{2|\Omega_e||y|}$$

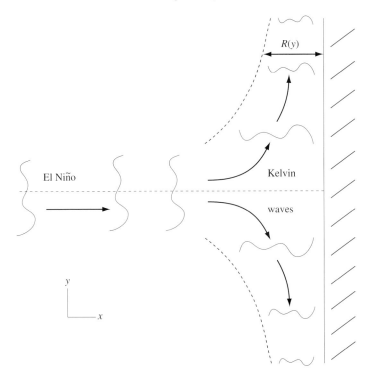

Fig. 4.7. The generation of poleward-propagating Kelvin waves along an eastern oceanic boundary by equatorial variability (e.g., during an El Niño event). The thick arrows indicate the propagation directions, and the thin wiggly lines indicate the width of the wave pattern in the perpendicular direction. The deformation radius, $R(y) = \sqrt{gH}/f(y)$, decreases with latitude since f increases, and the off-shore decay scale (i.e., pattern width) for the Kelvin waves decreases proportionally.

(since $f(y) \approx 2|\Omega_e|y/a$ near the Equator at $y = 0$ by (2.88), and N and H vary less with latitude than f does). The Kelvin-wave boundary-layer width shrinks as $|y|$ increases away from the Equator. This shallow-water equations interpretation is the one sketched in Fig. 4.4 with H the pycnocline depth.

   In addition to the extra-tropical modes, with $f_0 \neq 0$, analyzed in this section, there are analogous *equatorial* inertial, gravity, geostrophic, and Kelvin wave modes based on the equatorial β-plane approximation, $f \approx \beta_0 y$, where $y = 0$ is the Equator (Gill, 1982, Chapter 11). These other equatorial wave modes also have important roles in the El Niño scenario in both the atmosphere and ocean.

4.3 Geostrophic adjustment

The process called *geostrophic adjustment* is how a spatially localized, but otherwise arbitrary, initial condition in a rotating, stratified fluid evolves toward a localized flow that satisfies a diagnostic momentum balance (geostrophic, if $Ro \ll 1$) while radiating inertia-gravity waves away to distant regions. In general, geostrophic adjustment might be investigated by any of the following approaches.

(1) Solve an initial-value problem for the partial differential equation system, either analytically or numerically, and obtain an answer that confirms the phenomenological behavior described above.
(2) For the linear, conservative dynamics in (4.26), expand the initial state in the complete set of normal modes, and discard all but the geostrophic modes to represent the local *end state* of the adjustment process after all the inertia-gravity waves have propagated away.
(3) For the more general nonlinear, conservative dynamics in (4.5) and (4.8) with $\mathbf{F} = 0$, calculate the end state directly from the initial state by assuming Lagrangian conservation for the appropriate parcel invariants, assuming that the parcels remain in the neighborhood of their initial position (i.e., they are locally rearranged during geostrophic adjustment and are not carried away with the waves).

This third approach, originally taken by Carl Rossby, is the most general and least laborious way to determine the end state without having to keep track of the time evolution toward it. This approach is now illustrated for a simple situation where both the initial and final states are independent of the y coordinate and f is a constant. In this case the conservative, flat-bottom, shallow-water equations are

$$\frac{Du}{Dt} - fv = -g\frac{\partial\eta}{\partial x}$$

$$\frac{Dv}{Dt} + fu = 0$$

$$\frac{D\eta}{Dt} + (H + \eta)\frac{\partial u}{\partial x} = 0, \qquad (4.51)$$

with the substantial derivative having only 1D advection,

$$\frac{D}{Dt} = \frac{\partial}{\partial t} + u\frac{\partial}{\partial x}.$$

By defining $X(t)$ as the x coordinate for a Lagrangian parcel, then the following parcel invariants can be derived assuming that the velocity vanishes at $x = \pm\infty$.

Mass

$$M[X(t)] = \int^{X(t)} h\, dx', \qquad (4.52)$$

since

$$\frac{DM}{Dt} = \frac{dX}{dt}h + \int^{X(t)} \frac{\partial h}{\partial t}dx'$$

$$= uh + \int^{X(t)} \left[-\frac{\partial}{\partial x}(uh) \right] dx'$$

$$= uh - uh = 0. \tag{4.53}$$

Absolute momentum

$$A[X(t)] = fX + v, \tag{4.54}$$

since

$$\frac{DA}{Dt} = f\frac{dX}{dt} + \frac{Dv}{Dt}$$

$$= fu - fu = 0. \tag{4.55}$$

Absolute momentum in a parallel flow ($\partial_y = 0$) is the analog of absolute angular momentum, $A = \frac{1}{2}fr^2 + Vr$, in an axisymmetric flow ($\partial_\theta = 0$; cf., (3.83)).

Potential vorticity

$$Q[X(t)] = \frac{f + \partial_x v}{H_0 + \eta}, \tag{4.56}$$

from (4.24).

The other parcel invariants that are functionally related to these primary ones (e.g., Q^n for any n from (4.25)) are redundant with (4.52)–(4.54) and exert no further constraints on the parcel motion. In fact, this set of three parcel conservation relations is internally redundant by one relation since

$$Q = \frac{dA}{dX} \bigg/ \frac{dM}{dX}, \tag{4.57}$$

so only two of them are needed to fully determine the end state by the third approach among those listed at the start of this section. Which two is an option that may be chosen for analytical convenience.

Define a *parcel displacement* field by

$$\xi(t) = X(t) - X(0). \tag{4.58}$$

This allows the parcel invariance relations to be expressed as

$$M[X(t)] = M[X(0)] = M[X(t) - \xi(t)]$$
$$A[X(t)] = A[X(0)] = A[X - \xi]$$
$$Q[X(t)] = Q[X(0)] = Q[X - \xi] \tag{4.59}$$

for all t. In particular, make the hypothesis that the end state, at $t = \infty$, is a steady, geostrophically balanced one on all the parcels that both start and end in the vicinity of the initial disturbance. Thus, for the end state,

$$fv = g\frac{\partial \eta}{\partial x}. \tag{4.60}$$

Together the relations (4.59) and (4.60) suffice for calculating the end state, without having to calculate the intervening evolution that is usually quite complicated as parcels move around and inertia-gravity waves oscillate and radiate into the far-field.

A particular example is a local ridge at rest at $t = 0$ (Fig. 4.8):

$$\mathbf{u} = (0,0), \quad \eta = (\eta_0, 0) \quad \text{for} \quad (|x| < a_0, |x| > a_0), \tag{4.61}$$

where the parenthetical notation here and below indicates the inner (i) and outer (o) regional expressions in the format of (i, o). The symmetry of the initial condition about $x = 0$ is preserved under evolution. So only the half-space, $x \geq 0$, needs

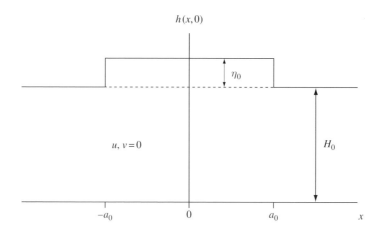

Fig. 4.8. An unbalanced ridge in the sea level elevation, $\eta(x)$, at $t = 0$.

to be considered. This initial condition has a particularly simple Q distribution, namely, piecewise constant:

$$Q = \left(Q_i = \frac{f}{H_0 + \eta_0}, \, Q_o = \frac{f}{H_0} \right).$$

For the end state the potential-vorticity parcel invariance (4.59) implies that

$$f + \frac{\partial v}{\partial x} = Q_{i,o}(H_0 + \eta) \tag{4.62}$$

in the two regions; the subscript indicates the relevant region. The boundary between the inner and outer regions is located at $x = a_\infty$, corresponding to the $X(t = \infty)$ value for the parcel with $X(0) = a_0$. Differentiate (4.62) with respect to x and substitute from (4.60) to obtain

$$\frac{\partial^2 v}{\partial x^2} - R_{i,o}^{-2} v = 0, \tag{4.63}$$

where

$$R_{i,o} = \sqrt{gH_{i,o}}/f$$

is the local deformation radius. The independent, homogeneous solutions for (4.63) are exponential functions, $e^{\pm x/R_{i,o}}$. Take the linear combination of the independent solutions in each region that satisfies the following boundary conditions:

$$v \to 0 \text{ at } x \to 0, \infty; \quad v \text{ continuous at } x = a_\infty. \tag{4.64}$$

These conditions are based on the odd symmetry of v relative to the point $x = 0$ (i.e., $v(x) = -v(-x)$, related by (4.60) to the even symmetry of η); spatial localization of the end-state flow; and continuity of v and η for all x. The result is

$$v = C \left(\sinh[x/R_i], \, \sinh[a_\infty/R_i]e^{-(x-a_\infty)/R_o} \right), \tag{4.65}$$

and, from (4.62),

$$\eta = \frac{Cf}{g} \left(R_i \cosh[x/R_i], \, -R_o \sinh[a_\infty/R_i]e^{-(x-a_\infty)/R_o} \right) + (\eta_0, 0). \tag{4.66}$$

Imposing continuity in η at $x = a_\infty$ yields

$$C = -\frac{g\eta_0}{f}[R_i \cosh[a_\infty/R_i] + R_o \sinh[a_\infty/R_i]]^{-1}. \tag{4.67}$$

These expressions are perhaps somewhat complicated to visualize. They become much simpler for the case of a wide ridge, where a_0, $a_\infty \gg R_i$, R_o. In this case (4.65)–(4.67) become

$$v = -\frac{g\eta_0}{f(R_i + R_o)}\left(e^{(x-a_\infty)/R_i}, e^{-(x-a_\infty)/R_o}\right)$$

$$\eta = \eta_0\left(1 - \frac{R_i}{R_i + R_o}e^{(x-a_\infty)/R_i}, \frac{R_o}{R_i + R_o}e^{-(x-a_\infty)/R_o}\right). \qquad (4.68)$$

Thus all the flow activity in the end state is in the neighborhood of the boundary between the inner and outer regions, and it is confined within a distance $\mathcal{O}(R_{o,i})$.

The only undetermined quantity in (4.65)–(4.68) is a_∞. It is related to ξ by

$$X(\infty) = a_\infty = a_0 + \xi(a_\infty). \qquad (4.69)$$

From (4.54) and (4.59),

$$fX(\infty) + v(X(\infty)) = fX(0)$$

$$\implies fx + v = f(x - \xi)$$

$$\implies \xi = -\frac{v}{f}. \qquad (4.70)$$

Inserting (4.65)–(4.67) into (4.70) and evaluating (4.69) yields an implicit equation for a_∞:

$$a_\infty = a_0 + \frac{\eta_0}{H_0}R_o\mathcal{J}_1, \qquad (4.71)$$

where the general expression for $\mathcal{J}_1$ is

$$\mathcal{J}_1 = \left[1 + \frac{R_i}{R_o\tanh[a_\infty/R_i]}\right]^{-1}. \qquad (4.72)$$

Equations (4.71) and (4.72) are somewhat complicated to interpret, but, for the case of small initial disturbances (i.e., $\eta_0/H_0 \ll 1$), they are simpler because $R_i \approx R_o$, $a_\infty \approx a_0$, and

$$\mathcal{J}_1 \approx [1 + 1/\tanh[a_0/R_o]]^{-1}. \qquad (4.73)$$

The function $\mathcal{J}_1 \approx a_0/R_o$ as $a_0/R_o \to 0$, and it approaches $1/2$ as a_0/R_o becomes large. Therefore,

$$a_\infty \approx a_0\left(1 + \frac{\eta_0}{H}\right),$$

and

$$a_\infty \approx a_0 + \frac{R_0}{2}\frac{\eta_0}{H}$$

in the respective limits. Note that (4.73) does not depend on a_∞, so the implicitness in (4.71) is resolved using the small-disturbance approximation. Also,

$$\xi = -\frac{C}{f}\left(\sinh[x/R_i], \ \sinh[a_\infty/R_i]e^{-(x-a_\infty)/R_0}\right), \tag{4.74}$$

with

$$C \approx \frac{\eta_0 N_0}{\sinh[a_0/R_0]}\mathcal{J}_1[a_0/R_0]$$

and $N_0 = \sqrt{g/H_0}$ from (4.40). So $C \approx \eta_0 N_0$ and $C \approx \eta_0 N_0 \exp[-a_0/R_0]$ in the respective limits, with the latter value a much smaller one for a wide ridge.

For a wide ridge ($a \gg R$), but not necessarily a small η_0/H value,

$$\mathcal{J}_1 = \frac{R_0}{R_i + R_0}, \tag{4.75}$$

and (4.71) and (4.74) become

$$a_\infty = a_0 + \left(\frac{\eta_0}{H}\right)\left(\frac{R_0}{R_i + R_0}\right)R_0 \tag{4.76}$$

$$\xi = \frac{g\eta_0}{f^2(R_i + R_0)}\left(e^{(x-a_\infty)/R_i}, \ e^{-(x-a_\infty)/R_0}\right). \tag{4.77}$$

Again the action is centered on the boundary within a distance $\mathcal{O}(R_{0,i})$. Equation (4.76) implies that the boundary itself moves a distance $\mathcal{O}(R_{0,i})$ under adjustment. This characteristic distance of deformation of the initial surface elevation profile under adjustment is why Rossby originally called R the deformation radius.

The end-state shapes for v, η, and ξ from (4.65)–(4.67) and (4.74) are shown in Fig. 4.9. $\eta(x)$ monotonically decays from the origin, and thus it remains a ridge as in the initial condition (NB, (4.61) and Fig. 4.8). However, its height is reduced, and its spatial extent is larger (i.e., it has slumped under the action of gravity). The parcel displacement is zero at the origin – what could determine whether a parcel at the center of the symmetric ridge goes east or west? The displacement field reaches a maximum at the potential vorticity boundary, $x = a_\infty > a_0$, and it decays away to infinity on the deformation radius scale, R_0. The velocity is anticyclonic inside $x = a_\infty$. At this location it reaches a maximum. Outside this boundary location, the vorticity is cyclonic, and the flow decays to zero at large x.

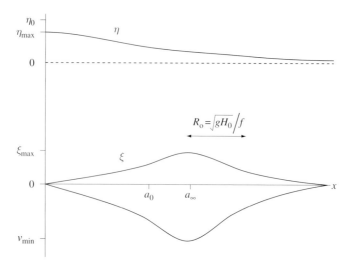

Fig. 4.9. A balanced end state ($t = \infty$) for the ridge after geostrophic adjustment: (top) sea level anomaly, $\eta(x)$; zonal parcel displacement, $\xi(x)$; and (bottom) meridional velocity, $v(x)$. a_0 and a_∞ are the initial and final locations of the parcel at the ridge edge (cf., Fig. 4.8).

Now estimate the end-state amplitudes, again by making the small-disturbance approximation, $\eta_0 \ll H_0$:

$$v_{\min} = -\eta_0 N_0 \mathcal{J}_1, \qquad \eta_{\max} = \eta_0 \mathcal{J}_2,$$

$$\xi_{\max} = \frac{\eta_0 R_0}{H_0} \mathcal{J}_1, \qquad a_\infty = a_0 + \xi_{\max}, \tag{4.78}$$

where

$$\mathcal{J}_2 = 1 - \frac{\mathcal{J}_1}{\sinh[a_\infty / R_i]} \approx 1 - \frac{\mathcal{J}_1}{\sinh[a_0 / R_0]}. \tag{4.79}$$

$\mathcal{J}_2 \approx a_0 / R_0$ and $\to 1$, respectively, for small and large a_0 / R_0 values.

For a wide ridge, $\mathcal{J}_2 = 1$, and (4.78) becomes

$$v_{\min} = -\eta_0 N_0 \left(\frac{R_0}{R_i + R_0} \right), \qquad \eta_{\max} = \eta_0,$$

$$\xi_{\max} = \frac{\eta_0 R_0}{H_0} \left(\frac{R_0}{R_i + R_0} \right), \qquad a_\infty = a_0 + \xi_{\max}. \tag{4.80}$$

The behavior is quite different for wide and narrow ridges, $a_0 \gg R_0$ and $a_0 \ll R_0$. For wide ridges $\eta_{\max} \approx \eta_0$ (indicating only a small amount of slumping); $v_{\min} \approx -R_0 f/2 \, \eta_0 / H_0$; and $\xi_{\max} \approx R_0 \eta_0 / 2H_0$ is only a small fraction of the initial ridge width. For the end state of wide ridges, v has adjusted to match $\eta(x, 0)$. Alternatively, for small-scale ridges, $\eta_{\max} \approx \eta_0 a_0 / R_o \ll \eta_0$; $v_{\min} \approx -a_0 f/2 \; \eta_0/H_0$; and $\xi_{\max} \approx a_0 \eta_0 / H_0$ (indicating a relatively large change from the initial shape with a big change in $\eta(x)$). In an alternative problem with an initially unbalanced, small-scale velocity patch, $v(x, 0)$, the result would be that η changes through geostrophic adjustment to match v.

Now analyze the energetics for geostrophic adjustment (4.17). The initial energy is entirely in the form of potential energy. The available potential energy per unit y length is

$$E_0 = \frac{1}{2} \int dx \, g \, \eta(x, 0)^2 = \frac{g}{2} \eta_0^2 a_0. \qquad (4.81)$$

An order-of-magnitude estimate for the local end-state energy after the adjustment is

$$E_\infty = \frac{1}{2} \int dx \left(h \mathbf{u}^2 + g \eta^2 \right) \sim \frac{g}{2} \eta_0^2 a_0 \left(\mathcal{J}_1^2 + \mathcal{J}_2^2 \right), \qquad (4.82)$$

using the magnitudes in (4.78) to make the estimate (NB, the detailed integration to evaluate E_∞ is a lengthy calculation). The kinetic and potential energies for the end state are of the same order, since $\mathcal{J}_1 \sim \mathcal{J}_2$. The ratio of final to initial energies is small for narrow ridges (with $\mathcal{J}_1, \mathcal{J}_2 \ll 1$), consistent with a large local change in the ridge shape and an inertia-gravity wave radiation away of the majority of the initial energy. However, for wide ridges (with $\mathcal{J}_1, \mathcal{J}_2 \sim 1$), the ratio is $\mathcal{O}(1)$, due to a relatively small amount of both changed ridge shape and radiated wave energy. In the limit with vanishing ridge size – a limit where the initial disturbance is too small to be affected by rotation – the initial disturbance leaves no local residue, and all of the initial energy goes into waves that propagate away from the disturbance site. (Think of a diver into a swimming pool.)

To further develop a qualitative understanding of the geostrophic adjustment process, Fig. 4.10 gives some 3D examples of the end states that result from motionless initial states with non-horizontal interfaces between fluid layers with different densities. In each configuration gravitational slumping and balanced flow development arise in patterns dictated by the initial density structure.

For more general circumstances, with $\partial_y \neq 0$, $\mathcal{F} \neq 0$, and $f \neq f_0$, an unbalanced state will not evolve through geostrophic adjustment into an exactly stationary end state. Nevertheless, insofar as $Ro, Fr < 1$ and the non-conservative rates are also slow, geostrophic adjustment still occurs. On a time scale of the wave radiation, $\sim 1/f$ or faster, inertia-gravity waves radiate away from the initial disturbance, leaving behind a geostrophic or gradient-wind balanced state. This latter flow pattern will subsequently evolve at advective and non-conservative rates, instead of being in a stationary state.

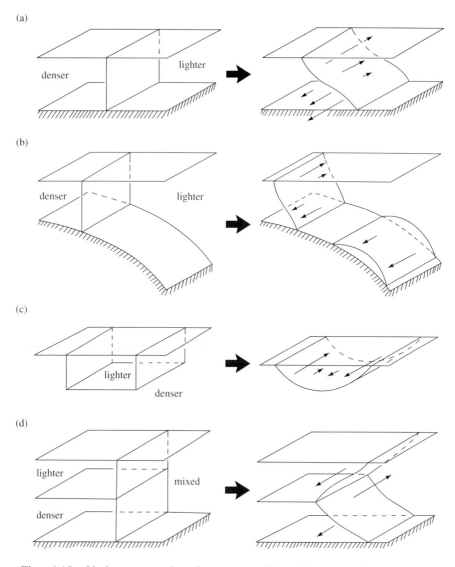

Fig. 4.10. Various examples for geostrophic adjustment. In each case the initial configuration (left column) has constant density layers with different density values, separated by the interfaces sketched here. The balanced, end-state, interface configurations are sketched in the right column. (Cushman-Roisin, 1994, Fig. 13-4.)

4.4 Gravity wave steepening: bores and breakers

Thus far only linear wave dynamics has been considered. In nature the inertia-gravity wave spectrum is typically broadly distributed in $\mathbf{k}$, and this breadth is maintained by nonlinear coupling among the different $\mathbf{k}$ components. A common assumption is that this coupling is weak, in the sense that the time scale for energy exchange among the components is longer than the time scale for their propagation. This assumption can be expressed either as $V/c \ll 1$ (weak flow), where V is the parcel velocity associated with the wave and c is the propagation speed, or as $\epsilon = ak \ll 1$ (small steepness), where a is the amplitude for a parcel displacement during a wave cycle and k is the wavenumber. A theoretical analysis based on an expansion in ϵ is called a *weak interaction theory*.

Yet another nonlinear behavior for gravity and other wave types is a *solitary wave*, i.e., a shape-preserving, spatially localized, uniformly propagating, solution of the conservative nonlinear dynamics, often arising through canceling tendencies between linear dispersive spreading and nonlinear steepening of the wave shape (e.g., the progression towards breaking, discussed in the next paragraph). In instances where a solitary-wave solution is conspicuously robust to perturbations, it is called a *soliton*. The shallow-water equations have no solitary wave solutions for an otherwise quiescent, uniform environment, but there are many known examples of solitary waves and solitons when the strict hydrostatic assumption is relaxed (e.g., the Korteweg–deVries approximation for long, weakly non-hydrostatic gravity waves in a uniform density fluid with a free surface; Whitham, 1999). The internal gravity wave shown in Fig. 4.2 is interpretable as a solitary wave.

Nonlinear dynamical effects in waves are not always weak and slow. This is most evident in the phenomenon of gravity *wave breaking* when the wave crest overtakes the trough and spills downward into it. The spilling phase typically fragments the interface – either the oceanic surface or an isopycnal surface in the stably stratified interior – and initiates turbulence, mixing, dissipation, and transfer (also known as deposition) of the wave momentum into more persistent winds or currents. For example, surface wave breaking is usually a very important step in the momentum transfer from the wind to the currents. The wind stress on the ocean acts primarily to generate surface waves that subsequently evolve until they break. Wave nonlinearity is also sometimes manifested in gravity *bores* that are sharp steps in the interfacial elevation that approximately maintain their shape as they propagate, usually accompanied by some amount of local turbulence and mixing. Tidal bores commonly occur at special sites, e.g., the Bay of Fundy along the Atlantic coast of Canada or the northeast corner of Brittany (near Mont St. Michel). They also occur on the downhill side of a *gravity current* caused by heavy fluid flowing downhill on a sloping surface, e.g., as induced by evening radiative cooling on a mountain slope.

The underlying cause for breakers and bores is the tendency for gravity waves to steepen on the forward-facing interface slope due to a faster local propagation speed when the elevation is higher, so that a high-elevation region will overtake a low-elevation one. In Section 4.2.2, the linear gravity wave speed is $c = \sqrt{gH}$. It is at least heuristically plausible that a locally larger $h = H + \eta$ might lead to a locally larger c. The classical analysis for this steepening behavior comes from compressible gas dynamics where the local speed of sound increases with the density variations within the wave. The acoustic outcome from such wave steepening is a *shock wave*, whereas the usual gravity wave outcome is breaking.

Assume 1D motions in the conservative, non-rotating, shallow-water equations above a flat bottom (NB, these assumptions preclude geostrophic flow and geostrophic adjustment). The governing equations are

$$\frac{\partial u}{\partial t} + u\frac{\partial u}{\partial x} + g\frac{\partial h}{\partial x} = 0$$

$$\frac{\partial h}{\partial t} + \frac{\partial}{\partial x}[uh] = 0. \tag{4.83}$$

The momentum and thickness equations can be combined into a first-order wave equation,

$$\frac{\partial \gamma}{\partial t} + V\frac{\partial \gamma}{\partial x} = 0, \tag{4.84}$$

with two separate definitions for the composite variable, γ (called the *Reimann invariant*) and propagation velocity, V (called the *characteristic velocity*):

$$\gamma_\pm = u \pm 2\sqrt{gh}, \quad V_\pm = u \pm \sqrt{gh}. \tag{4.85}$$

This can be verified by substituting (4.85) into (4.84) and by using (4.83) to evaluate the time derivatives of u and h. The characteristic equation (4.84) has a general solution, $\gamma = \Gamma(\xi)$, for the composite coordinate, $\xi(x,t)$ (called the *characteristic coordinate*) defined implicitly by

$$\xi + V(\xi)t = x. \tag{4.86}$$

The demonstration that this is a solution comes from taking the t and x derivatives of (4.86),

$$\partial_t\xi + t\, d_\xi V\, \partial_t\xi + V = 0, \quad \partial_x\xi + t\, d_\xi V\, \partial_x\xi = 1$$

(with $d_\xi V = dV/d\xi$); solving for $\partial_t\xi$ and $\partial_x\xi$; substituting them into expressions for the derivatives of γ,

$$\partial_t\gamma = \partial_\xi\Gamma\, \partial_t\xi = -\frac{\partial_\xi\Gamma}{1+d_\xi V\, t}V, \quad \partial_x\gamma = \partial_\xi\Gamma\, \partial_x\xi = \frac{\partial_\xi\Gamma}{1+d_\xi V\, t};$$

and finally inserting the latter into (4.84),

$$\partial_t\gamma + V\partial_x\gamma = \frac{\partial_\xi\Gamma}{1+d_\xi V\, t}(-V+V) = 0.$$

The function Γ is determined by an initial condition,

$$\xi(x,0) = x, \quad \Gamma(\xi) = \gamma(x,0).$$

Going forward in time, γ preserves its initial value, $\Gamma(\xi)$, but this value moves to a new location, $X(\xi, t) = \xi + V(\xi)t$, by propagating at a speed $V(\xi)$. The speed V is $\approx \pm\sqrt{gH}$ after neglecting the velocity and height departures, $(u, h - H)$, from the resting state in (4.85); these approximate values for V are the familiar linear gravity wave speeds for equal and oppositely directed propagation (Section 4.2.2). When the fluctuation amplitudes are not negligible, then the propagation speeds differ from the linear speeds and are spatially inhomogeneous.

In general two initial conditions must be specified for the second-order partial differential equation system (4.83). This is accomplished by specifying conditions for $\gamma_+(x)$ and $\gamma_-(x)$ that then have independent solutions, $\gamma_\pm(\xi_\pm)$. A particular solution for propagation in the $+\hat{\mathbf{x}}$ direction is

$$\Gamma_- = -2\sqrt{gH}, \quad \Gamma_+ = 2\sqrt{gH} + \delta\Gamma,$$

$$\delta\Gamma(\xi_+) = 4\left(\sqrt{g\mathcal{H}} - \sqrt{gH}\right),$$

$$h(X, t) = \mathcal{H}(\xi_+), \quad u(X, t) = 2\left(\sqrt{g\mathcal{H}} - \sqrt{gH}\right),$$

$$X(\xi_+, t) = \xi_+ + V_+(\xi_+)t, \quad V_+ = \left(3\sqrt{g\mathcal{H}} - 2\sqrt{gH}\right), \tag{4.87}$$

where $\mathcal{H}(x) = H + \eta(x) > 0$ is the initial layer thickness shape. Here $V_+ > 0$ whenever $\mathcal{H} > 4/9\,H$, and both V and u increase with increasing $\mathcal{H}$. When h is larger, $X(t)$ progresses faster and vice versa. For an isolated wave of elevation (Fig. 4.11), the characteristics converge on the forward side of the wave and diverge on the backward side. This leads to a steepening of the front of the wave form and a reduction of its slope in the back. Since V is constant on each characteristic, these tendencies are inexorable; therefore, at some time and place a characteristic on the forward face will catch up with another one ahead of it. Beyond this point the solution will become multi-valued in γ, h, and u and thus invalidate the shallow-water equations assumptions. This situation can be interpreted as the possible onset for a wave breaking event, whose accurate description requires more general dynamics than the shallow-water equations.

An alternative interpretation is that a collection of intersecting characteristics may create a discontinuity in h (i.e., a downward step in the propagation direction) that can then continue to propagate as a generalized shallow-water equations solution (Fig. 4.12). In this interpretation the solution is a bore, analogous to a shock. *Jump conditions* for the discontinuities in (u, h) across the bore are derived from the governing equations (4.83) expressed in flux-conservation form, namely,

$$\frac{\partial p}{\partial t} + \frac{\partial q}{\partial x} = 0$$

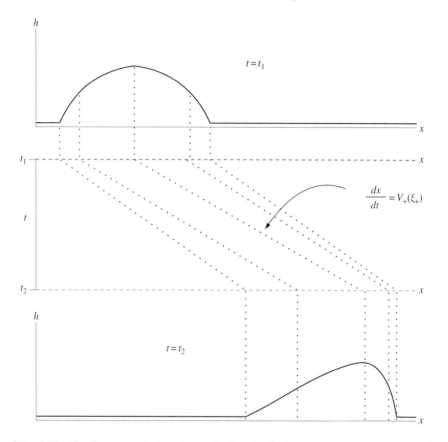

Fig. 4.11. Nonlinear evolution for an isolated, shallow-water, gravity wave of elevation. The wave shape at the earlier time ($t = t_1$; top) evolves into the shape at a later time ($t = t_2$; bottom) that has a shallower slope on its backward face and a steeper slope on its forward face. This example is for a rightward propagating wave. The characteristic coordinate, ξ_+, remains constant for each point on the wave, but its associated velocity, V_+, is larger where the elevation is higher (shown by the line slopes in the middle diagram).

for some "density," p, and "flux," q. (The thickness equation is already in this form with $p = h$ and $q = hu$. The momentum equation may be combined with the thickness equation to give a second flux-conservation equation with $p = hu$ and $q = hu^2 + gh^2/2$.) This equation type has the integral interpretation that the total amount of p between any two points, $x_1 < x_2$, can only change due to the difference in fluxes across these points:

$$\frac{d}{dt}\int_{x_1}^{x_2} p\,dx = -(q_2 - q_1).$$

Now assume that p and q are continuous on either side of a discontinuity at $x = X(t)$ that itself moves with speed, $U = dX/dt$. At any instant define

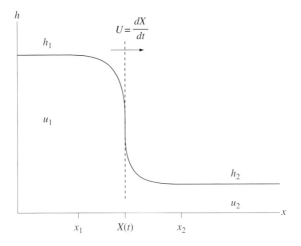

Fig. 4.12. A gravity bore, with a discontinuity in (u, h). The bore is at $x = X(t)$, and it moves with a speed, $u = U(t) = dX/dt$. Subscripts 1 and 2 refer to locations to the left and right of the bore, respectively.

neighboring points, $x_1 > X > x_2$. The left-hand side can be evaluated as

$$
\frac{d}{dt} \int_{x_1}^{x_2} p \, dx = \frac{d}{dt} \left(\int_{x_1}^{X} p \, dx + \int_{X}^{x_2} p \, dx \right)
$$
$$
= p(X^-)\frac{dX}{dt} - p(X^+)\frac{dX}{dt}
$$
$$
+ \left(\int_{x_1}^{X} \frac{\partial p}{\partial t} \, dx + \int_{X}^{x_2} \frac{\partial p}{\partial t} \, dx \right). \tag{4.88}
$$

The $-$ and $+$ superscripts for X indicate values on the left and right of the discontinuity. Now take the limit as $x_1 \rightarrow X^-$ and $x_2 \rightarrow X^+$. The final integrals vanish since $|x_2 - X|$, $|X - x_1| \rightarrow 0$ and p_t is bounded in each of the sub-intervals. Thus,

$$
U\Delta[p] = \Delta[q],
$$

and $\Delta[a] = a(X^+) - a(X^-)$ denotes the difference in values across the discontinuity. For this bore problem this type of analysis gives the following jump conditions for the mass and momentum:

$$
U\Delta h = \Delta[uh], \quad U\Delta[uh] = \Delta[hu^2 + gh^2/2]. \tag{4.89}
$$

For given values of $h_1 > h_2$ and u_2 (the wave velocity at the overtaken point), the bore propagation speed is

$$U = u_2 + \sqrt{\frac{gh_1(h_1 + h_2)}{2h_2}} > u_2, \tag{4.90}$$

and the velocity behind the bore is

$$u_1 = u_2 + \frac{h_1 - h_2}{h_1}\sqrt{\frac{gh_1(h_1 + h_2)}{2h_2}} > u_2 \text{ and } < U. \tag{4.91}$$

The bore propagates faster than the fluid velocity on either side of the discontinuity.

For further analysis of this and many other nonlinear wave problems see Whitham (1999).

4.5 Stokes drift and material transport

From (4.28) and (4.29) the shallow-water equations inertia-gravity wave in (4.37) has an eigensolution form,

$$\eta = \eta_0 \cos[\Theta]$$

$$\mathbf{u}_{\text{h}} = \frac{g\eta_0}{gHK^2}\left(\omega\mathbf{k}\cos[\Theta] - f\hat{\mathbf{z}} \times \mathbf{k}\sin[\Theta]\right)$$

$$w = \frac{\omega\eta_0 z}{H}\sin[\Theta], \tag{4.92}$$

where $\Theta = \mathbf{k} \cdot \mathbf{x} - \omega t$ is the wave phase function and η_0 is a real constant.

The time or wave-phase average of all these wave quantities is zero, e.g., the Eulerian mean velocity, $\bar{\mathbf{u}} = 0$. However, the average Lagrangian velocity is not zero for a trajectory in (4.92). To demonstrate this, decompose the trajectory, $\mathbf{r}$, into a "mean" component that uniformly translates and a wave component that oscillates with the wave phase:

$$\mathbf{r}(t) = \bar{\mathbf{r}}(t) + \mathbf{r}'(t), \tag{4.93}$$

where

$$\bar{\mathbf{r}} = \mathbf{u}^{\text{st}}\, t \tag{4.94}$$

and $\mathbf{u}^{\text{st}}$ is called the *Stokes drift* velocity. By definition any fluctuating quantity has a zero average over the wave phase. The formula for $\mathbf{u}^{\text{st}}$ is derived by making a Taylor series expansion of the trajectory equation (2.1) around the evolving mean position, $\bar{\mathbf{r}}$, then taking a wave-phase average:

$$\frac{d\mathbf{r}}{dt} = \mathbf{u}(\mathbf{r})$$

$$\frac{d\bar{\mathbf{r}}}{dt} + \frac{d\mathbf{r}'}{dt} = \mathbf{u}'(\bar{\mathbf{r}} + \mathbf{r}')$$

$$= \mathbf{u}'(\bar{\mathbf{r}}) + (\mathbf{r}' \cdot \nabla)\mathbf{u}'(\bar{\mathbf{r}}) + \mathcal{O}(\mathbf{r}'^2)$$

$$\implies \frac{d\bar{\mathbf{r}}}{dt} \approx \overline{(\mathbf{r}' \cdot \nabla)\mathbf{u}'} = \mathbf{u}^{\text{st}}. \tag{4.95}$$

(Here all vectors are 3D.) Further, a formal integration of the fluctuating trajectory equation yields the more common expression for Stokes drift, namely,

$$\mathbf{u}^{\text{st}} = \overline{\left(\left(\int^t \mathbf{u}'dt\right) \cdot \nabla\right)\mathbf{u}'}. \tag{4.96}$$

A non-zero Stokes drift is possible for any kind of fluctuation (cf. Section 5.3.5). The Stokes drift for an inertia-gravity wave is evaluated using (4.92) in (4.96). The fluctuating trajectory is

$$\mathbf{r}'_{\text{h}} = -\frac{g\eta_0}{gHK^2}\left(\mathbf{k}\sin[\Theta] + \frac{f}{\omega}\hat{\mathbf{z}} \times \mathbf{k}\cos[\Theta]\right),$$

$$r^{z'} = \frac{\eta_0 z}{H}\cos[\Theta]. \tag{4.97}$$

Since the phase averages of $\cos^2[\Theta]$ and $\sin^2[\Theta]$ are both equal to $1/2$ and the average of $\cos[\Theta]\sin[\Theta]$ is zero, the Stokes drift is

$$\mathbf{u}^{\text{st}} = \frac{1}{2}\frac{\omega}{(HK)^2}\eta_0^2\mathbf{k}\,. \tag{4.98}$$

The Stokes drift is purely a horizontal velocity, parallel to the wavenumber vector, $\mathbf{k}$, and the phase velocity, $\mathbf{c}_{\text{p}}$. It is small compared to $\mathbf{u}'_{\text{h}}$ since it has a quadratic dependence on η_0, rather than a linear one, and the wave modes are derived with the linearization approximation that $\eta_0/H \ll 1$. The mechanism behind Stokes drift is the following: when a wave-induced parcel displacement, $\mathbf{r}'$, is in the direction of propagation, the wave pattern movement sustains the time interval when the wave velocity fluctuation, $\mathbf{u}'$ is in that same direction; whereas, when the displacement is opposite to the pattern propagation direction, the advecting wave velocity is more briefly sustained. Averaging over a wave cycle, there is net motion in the direction of propagation.

Stokes drift is essentially due to the gravity-wave rather than inertia-wave behavior. In the short-wave limit (i.e., gravity waves; Section 4.2.2), (4.98) becomes

$$\mathbf{u}^{\text{st}} \rightarrow \sqrt{\frac{g}{H}} \frac{\eta_0^2}{H} \left(\frac{\mathbf{k}}{K} \right) = \frac{u_{\text{h0}}^2}{2\sqrt{gH}} \left(\frac{\mathbf{k}}{K} \right),$$

where $u_{\text{h0}} \rightarrow \sqrt{g/H}\,\eta_0$ is the horizontal velocity amplitude of the eigensolution in (4.92). These expressions are independent of K, and $\mathbf{u}^{\text{st}}$ has a finite value. In the long-wave limit (i.e., inertia-waves), (4.98) becomes

$$\mathbf{u}^{\text{st}} \rightarrow \frac{f\eta_0^2}{2H^2 K} \left(\frac{\mathbf{k}}{K} \right) = \frac{K u_{\text{h0}}^2}{2f} \left(\frac{\mathbf{k}}{K} \right),$$

with $u_{\text{h0}} \rightarrow (f/HK)\,\eta_0$ from (4.92). This shows that $\mathbf{u}^{\text{st}} \rightarrow 0$ as $K \rightarrow 0$ in association with finite u_{h0} and vanishing η_0 (and w_0); i.e., because inertial oscillations have a finite horizontal velocity and vanishing free-surface displacement and vertical velocity (Section 2.4.3), they induce no Stokes drift.

Stokes drift can be interpreted as a wave-induced mean mass flux (equivalent to a wave-induced fluid volume flux times ρ_0 for a uniform density fluid). Substituting

$$h = \overline{h} + h'$$

into the thickness equation (4.5) and averaging yields the following equation for the evolution of the wave-averaged thickness,

$$\frac{\partial \overline{h}}{\partial t} + \mathbf{\nabla}_{\text{h}} \cdot (\overline{h}\overline{\mathbf{u}}_{\text{h}}) = -\mathbf{\nabla}_{\text{h}} \cdot (\overline{h'\mathbf{u}_{\text{h}}'}), \tag{4.99}$$

that includes the divergence of *eddy mass flux*, $\rho_0 \overline{h'\mathbf{u}_{\text{h}}'}$. Since $h' = \eta'$ for the shallow-water equations, the inertia-gravity wave solution (4.92) implies that

$$\overline{h'\mathbf{u}_{\text{h}}'} = \frac{1}{2} \frac{H\omega}{(HK)^2} \eta_0^2 \mathbf{k} = H\mathbf{u}^{\text{st}}. \tag{4.100}$$

The depth-integrated Stokes transport, $\int_0^H \mathbf{u}^{\text{st}}\,dz$, is equal to the eddy mass flux.

A similar formal averaging of the shallow-water equations tracer equation for $\tau(\mathbf{x}_{\text{h}}, t)$ yields

$$\frac{\partial \overline{\tau}}{\partial t} + \overline{\mathbf{u}}_{\text{h}} \cdot \mathbf{\nabla}_{\text{h}}\overline{\tau} = -\overline{\mathbf{u}_{\text{h}}' \cdot \mathbf{\nabla}_{\text{h}}\tau'}. \tag{4.101}$$

If there is a large-scale, "mean" tracer field, $\overline{\tau}$, then the wave motion induces a tracer fluctuation,

$$\frac{\partial \tau'}{\partial t} \approx -\mathbf{u}_{\text{h}}' \cdot \mathbf{\nabla}_{\text{h}}\overline{\tau}$$

$$\implies \tau' \approx -\left(\int^t \mathbf{u}_{\text{h}}'\,dt \right) \cdot \mathbf{\nabla}_{\text{h}}\overline{\tau}, \tag{4.102}$$

in a linearized approximation. Using this τ' plus $\mathbf{u}'_h$ from (4.92), the wave-averaged effect in (4.101) is evaluated as

$$-\overline{\mathbf{u}'_h \cdot \boldsymbol{\nabla}_h \tau'} = \overline{\mathbf{u}'_h \left(\int^t \mathbf{u}'_h \, dt \right) \cdot \boldsymbol{\nabla}_h \overline{\tau}}$$

$$\approx -\overline{\left(\left(\int^t \mathbf{u}'_h \, dt \right) \cdot \boldsymbol{\nabla}_h \right) \mathbf{u}'_h \cdot \boldsymbol{\nabla}_h \overline{\tau}}$$

$$= -\mathbf{u}^{st} \cdot \boldsymbol{\nabla}_h \overline{\tau}. \tag{4.103}$$

The step from the first line to the second involves an integration by parts in time, interpreting the averaging operator as a time integral over the rapidly varying wave phase, and neglecting the space and time derivatives of $\overline{\tau}(\mathbf{x}_h, t)$ compared to those of the wave fluctuations (i.e., the mean fields vary slowly compared to the wave fields). Inserting (4.103) into (4.101) yields the final form for the large-scale tracer evolution equation, namely,

$$\frac{\partial \tau}{\partial t} + \mathbf{u}_h \cdot \boldsymbol{\nabla}_h \tau = -\mathbf{u}^{st} \cdot \boldsymbol{\nabla}_h \tau, \tag{4.104}$$

where the overbar averaging symbols are now implicit. Thus, wave-averaged material concentrations are advected by the wave-induced Stokes drift in addition to their more familiar advection by the wave-averaged velocity.

A similar derivation yields a wave-averaged *vortex force* term proportional to $\mathbf{u}^{st}$ in the mean momentum equation. This vortex force is believed to be the mechanism for creating *wind rows*, or *Langmuir circulations*, which are convergence-line patterns in surface debris often observed on lakes or the ocean in the presence of surface gravity waves.

By comparison with the eddy-diffusion model (3.109), the eddy-induced advection by Stokes drift is a very different kind of eddy–mean interaction. The reason for this difference is the distinction between the random velocity assumed for eddy diffusion and the periodic wave velocity for Stokes drift.

4.6 Quasigeostrophy

The *quasigeostrophic approximation* for the shallow-water equations is an asymptotic approximation in the limit

$$Ro \rightarrow 0, \quad \mathcal{B} = (Ro/Fr)^2 = \mathcal{O}(1). \tag{4.105}$$

$\mathcal{B} = (NH/fL)^2 = (R/L)^2$ is the *Burger number*. Now make the shallow-water equations non-dimensional with a transformation of variables based on the following geostrophic scaling estimates:

$$x, y \sim L, \quad u, v \sim V,$$

$$h \sim H_0, \quad t \sim \frac{L}{V},$$

$$f \sim f_0, \quad p \sim \rho_0 V f_0 L,$$

$$\eta, B \sim \epsilon H_0, \quad \beta = \frac{df}{dy} \sim \epsilon \frac{f_0}{L},$$

$$w \sim \epsilon \frac{V H_0}{L}, \quad \mathbf{F} \sim \epsilon f_0 V. \tag{4.106}$$

$\epsilon \ll 1$ is the expansion parameter (e.g., $\epsilon = Ro$). Estimate the dimensional magnitude of the terms in the horizontal momentum equation as follows:

$$\frac{D\mathbf{u}}{Dt} \sim V^2/L = Ro f_0 V$$

$$f\hat{\mathbf{z}} \times \mathbf{u} \sim f_0 V$$

$$\frac{1}{\rho_0} \nabla p \sim f_0 VL/L = f_0 V$$

$$\mathbf{F} \sim \epsilon f_0 V_0. \tag{4.107}$$

Substitute for non-dimensional variables, e.g.,

$$\mathbf{x}_{\text{dim}} = L\, \mathbf{x}_{\text{non-dim}} \text{ and } \mathbf{u}_{\text{dim}} = V\, \mathbf{u}_{\text{non-dim}}, \tag{4.108}$$

and divide by $f_0 V$ to obtain the non-dimensional momentum equation,

$$\epsilon \frac{D\mathbf{u}}{Dt} + f\hat{\mathbf{z}} \times \mathbf{u} = -\nabla p + \epsilon \mathbf{F}$$

$$\frac{D}{Dt} = \frac{\partial}{\partial t} + \mathbf{u} \cdot \nabla$$

$$f = 1 + \epsilon \beta y. \tag{4.109}$$

These expressions are entirely in terms of non-dimensional variables, where from now on the subscripts in transformation formulae like (4.108) are deleted for brevity. A β-plane approximation has been made for the Coriolis frequency in (4.109). The additional non-dimensional relations for the shallow-water equations are

$$p = \mathcal{B}\eta, \quad h = 1 + \epsilon(\eta - B)$$

$$\epsilon \frac{\partial \eta}{\partial t} + \epsilon \nabla \cdot [(\eta - B)\mathbf{u}] = -\nabla \cdot \mathbf{u}. \tag{4.110}$$

Now investigate the quasigeostrophic limit (4.105) for (4.109) and (4.110) as $\epsilon \to 0$ with β, $\mathcal{B} \sim 1$. The leading order balances are

$$\hat{\mathbf{z}} \times \mathbf{u} = -\mathcal{B} \nabla \eta, \quad \nabla \cdot \mathbf{u} = 0 . \tag{4.111}$$

This in turn implies that the geostrophic velocity, **u**, can be approximately represented by a streamfunction, $\psi = \mathcal{B}\eta$. Since the geostrophic velocity is non-divergent, there is a divergent horizontal velocity component only at the next order of approximation in ϵ. A perturbation expansion is being made for all the dependent variables, e.g.,

$$\mathbf{u} = \hat{\mathbf{z}} \times \mathcal{B}\nabla\eta + \epsilon\,\mathbf{u}_{\mathrm{a}} + \mathcal{O}(\epsilon^2).$$

The $\mathcal{O}(\epsilon)$ component is called the *ageostrophic velocity*, $\mathbf{u}_{\mathrm{a}}$. The dimensional scale for $\mathbf{u}_{\mathrm{a}}$ is therefore ϵV. It joins with w (whose scale in (4.106) is similarly reduced by the factor of ϵ) in a 3D continuity balance at $\mathcal{O}(\epsilon V/L)$, namely,

$$\nabla\cdot\mathbf{u}_{\mathrm{a}} + \frac{\partial w}{\partial z} = 0. \tag{4.112}$$

The ageostrophic and vertical currents are thus much weaker than the geostrophic currents.

Equations (4.109)–(4.111) comprise an under-determined system, with three equations for four unknown dependent variables. To complete the quasi-geostrophic system, another relation must be found that is well ordered in ϵ. This extra relation is provided by the potential vorticity equation, as in (4.24) but here it is non-dimensional and approximated as $\epsilon \to 0$. The dimensional potential vorticity is scaled by f_0/H_0 and has the non-dimensional expansion,

$$q = 1 + \epsilon q_{\mathrm{QG}} + \mathcal{O}(\epsilon^2), \quad q_{\mathrm{QG}} = \nabla^2\psi - \mathcal{B}^{-1}\psi + \beta y + B. \tag{4.113}$$

Note that this potential vorticity contains contributions from both the motion (the relative and stretching vorticity terms) and the environment (the planetary and topographic terms). Its parcel conservation equation to leading order is

$$\left[\frac{\partial}{\partial t} + J[\psi,\ \ ]\right] q_{\mathrm{QG}} = \mathcal{F}, \tag{4.114}$$

where only the geostrophic velocity advection contributes to the conservative parcel rearrangements of q_{QG}. This relation completes the posing for the quasi-geostrophic dynamical system. Furthermore, it can be viewed as a single equation for ψ only (as was also true for the potential vorticity equation in a 2D flow; Section 3.1.2). Alternatively, the derivation of (4.113) and (4.114) can be performed directly by taking the curl of the horizontal momentum equation and combining it with the thickness equation, with due attention to the relevant order in ϵ for the contributing terms.

The energy equation in the quasigeostrophic limit is somewhat simpler than the general shallow-water equations relation (4.17). It is obtained by multiplying (4.114) by $-\psi$ and integrating over space. For conservative motions ($\mathcal{F} = 0$), the non-dimensional energy principle for quasigeostrophy is

$$\frac{dE}{dt} = 0, \quad E = \iint dx\,dy\, \frac{1}{2}\left[(\nabla\psi)^2 + \mathcal{B}^{-1}\psi^2\right].\tag{4.115}$$

The ratio of kinetic to available potential energy is on the order of $\mathcal{B}$; for $L \gg R$, most of the energy is potential, and vice versa.

The quasigeostrophic system is a first order partial differential equation in time, similar to the barotropic vorticity equation (3.30), whereas the shallow-water equations are third order (cf., (4.28)). This indicates that quasigeostrophy has only a single type of normal mode, rather than both geostrophic and inertia-gravity wave mode types as in the shallow-water equations (as well as the 3D primitive and Boussinesq equations). Under the conditions $\beta = B = 0$, the mode type retained by this approximation is the geostrophic mode, with $\omega = 0$. Generally, however, this mode has $\omega \neq 0$ when β and/or $B \neq 0$. By the scaling estimates (4.106),

$$\omega \sim \frac{V}{L} = \epsilon f_0.\tag{4.116}$$

Hence, any quasigeostrophic wave modes have a frequency $\mathcal{O}(\epsilon)$ smaller than the inertia-gravity modes that all have $|\omega| \geq f_0$. This supports the common characterization that the quasigeostrophic modes are *slow modes* and the inertia-gravity modes are *fast modes*. A related characterization is that balanced motions (e.g., quasigeostrophic motions) evolve on the *slow manifold* that is a sub-space of the possible solutions of the shallow-water equations (or primitive and Boussinesq equations).

The quasigeostrophic shallow-water equations model has analogous stationary states to the barotropic model (Section 3.1.4), namely, axisymmetric vortices when $f = f_0$ and zonal parallel flows for general $f(y)$ (and others that are not discussed here). The most important difference between barotropic and shallow-water equations stationary solutions is the more general definition for q in shallow-water equations. The quasigeostrophic model also has a $(\psi, y) \leftrightarrow (-\psi, -y)$ parity symmetry (cf., (3.52) in Section 3.1.4), although the general shallow-water equations do not. Thus, cyclonic and anticyclonic dynamics are fundamentally equivalent in quasigeostrophy (as in 2D; Section 3.1.2), but different in more general dynamical systems such as the shallow-water equations.

The non-dimensional shallow-water equations quasigeostrophic dynamical system (4.109)–(4.114) is alternatively but equivalently expressed in dimensional variables as follows:

$$p = g\rho_0\,\eta, \quad \psi = \frac{g}{f_0}\eta, \quad h = H + \eta - B, \quad f = f_0 + \beta_0 y,$$

$$\mathbf{u} = -\frac{g}{f_0}\hat{\mathbf{z}}\times\nabla\eta, \quad q_{\mathrm{QG}} = \nabla^2\psi - R^{-2}\psi + \beta_0 y + \frac{f_0}{H}B,$$

$$\left[\frac{\partial}{\partial t} + J[\psi,\ ]\right]q_{\mathrm{QG}} = \mathcal{F}. \tag{4.117}$$

These relations can be derived by reversing the non-dimensional transformation of variables in the preceding relations, or they could be derived directly from the dimensional shallow-water equations with appropriate approximations. The real value of non-dimensionalization in GFD is as a guide to consistent approximation. The non-dimensionalized derivation in (4.109)–(4.114) is guided by the perturbation expansion in $\epsilon \ll 1$. In contrast, ϵ does not appear in either the dimensional shallow-water equations (4.1)–(4.8) or quasigeostrophic (4.117) systems, so the approximate relation of the latter to the former is somewhat hidden.

4.7 Rossby waves

The archetype of a quasigeostrophic wave is a *planetary* or *Rossby wave* that arises from the approximately spherical shape of rotating Earth as manifested through $\beta \neq 0$. Quasigeostrophic wave modes can also arise from bottom slopes ($\nabla B \neq 0$) and are then called *topographic Rossby waves*. A planetary Rossby wave is illustrated by writing the quasigeostrophic system (4.113)–(4.115) linearized around a resting state:

$$\frac{\partial}{\partial t}[\nabla^2\psi - \mathcal{B}^{-1}\psi] + \beta\frac{\partial\psi}{\partial x} = 0. \tag{4.118}$$

For normal mode solutions with

$$\psi = \mathrm{Real}\left(\psi_0\,e^{i(\mathbf{k}\cdot\mathbf{x}-\omega t)}\right), \tag{4.119}$$

the Rossby-wave dispersion relation is

$$\left(-i\omega[-K^2 - \mathcal{B}^{-1}] + ik\beta\right)\psi_0 = 0$$

$$\implies \quad \omega = -\frac{\beta k}{K^2 + \mathcal{B}^{-1}}. \tag{4.120}$$

(For comparison – in the spirit of the dimensional quasigeostrophic relations (4.117) – an equivalent Rossby-wave dispersion relation is

$$\omega = -\frac{\beta_0 k}{K^2 + R^{-2}},$$

where all quantities here are dimensional.) The zonal phase speed (i.e., the velocity that its spatial patterns move with; Section 4.2) is westward everywhere since

$$\omega k < 0, \tag{4.121}$$

but its group velocity, the velocity for wave energy propagation, in non-dimensional form is

$$\mathbf{c}_g = \frac{\partial \omega}{\partial \mathbf{k}} = \left(\frac{\beta[k^2 - \ell^2 - \mathcal{B}^{-1}]}{[K^2 + \mathcal{B}^{-1}]^2}, \frac{2\beta k\ell}{[K^2 + \mathcal{B}^{-1}]^2} \right). \tag{4.122}$$

$\mathbf{c}_g$ can be oriented in any direction, depending upon the signs of ω, k, and ℓ. The long-wave limit ($K \to 0$) of (4.120) is non-dispersive,

$$\omega \to -\mathcal{B}\,\beta k, \tag{4.123}$$

and the associated group velocity must also be westward. To be within the long-wave limit, the distinguishing spatial scale is $K^{-1} = \mathcal{B}^{1/2}$, or, in dimensional terms, $L = R$. (Again note the significant role of the deformation radius.) For shorter waves, (4.120) is dispersive. If the wave is short enough and has a zonal orientation to its propagation, with

$$k^2 > \ell^2 + \mathcal{B}^{-1},$$

the zonal group velocity is eastward even though the phase propagation remains westward. In other words, only a Rossby wave shorter than the deformation radius can carry energy eastward.

Due to the scaling assumptions in (4.106) about β and B, the general shallow-water equations wave analysis could be redone for (4.109) and (4.110) with the result that only $\mathcal{O}(\epsilon)$ corrections to the f-plane inertia-gravity modes are needed. This more general analysis, however, would be significantly more complicated because the linear shallow-water equations (4.109) and (4.110) no longer have constant coefficients, and the normal mode solution forms are no longer the simple trigonometric functions in (4.29).

Further analyses of Rossby waves are in Pedlosky (Sections 3.9–26, 1987) and Gill (Sections 11.2–7, 1982).

4.8 Rossby-wave emission

An important purpose of GFD is the idealization and abstraction of the various physical influences causing a given phenomenon (Chapter 1). But an equally important, but logically subsequent, purpose is to deliberately combine influences to see what modifications arise in the resultant phenomena. Here consider two instances where simple f-plane solutions – an isolated, axisymmetric vortex

(Section 3.1.4), and a boundary Kelvin wave (Section 4.2.3) – lose their exact validity on the β-plane and consequently behave somewhat differently. In each case some of the energy in the primary phenomenon is converted into Rossby-wave energy through processes that can be called *wave emission* or *wave scattering*.

4.8.1 Vortex propagation on the β-plane

Assume an initial condition with an axisymmetric vortex in an unbounded domain on the β-plane with no non-conservative influences. Further assume that $Ro \ll 1$ so that the quasigeostrophic approximation (Section 4.6) is valid. If β were zero, the vortex would be a stationary state, and for certain velocity profiles, $V(r)$, it would be stable to small perturbations. However, for $\beta \neq 0$, no such axisymmetric stationary states can exist.

So what happens to such a vortex? In a general way, it seems plausible that it might not change much for a strong enough vortex. A scaling estimate for the ratio of the β term and vorticity advection in the potential vorticity equation (4.114) is

$$\mathcal{R} = \frac{\beta v}{\mathbf{u} \cdot \nabla \zeta} \sim \frac{\beta V}{V(1/L)(V/L)} = \frac{\beta L^2}{V} ; \qquad (4.124)$$

this must be small for the β influence to be weak. The opposite situation occurs when R is large. In this case the initially axisymmetric ψ pattern propagates westward and changes its shape by Rossby-wave dispersion (when L/R is not large).

A numerical solution of (4.114) for an anticyclonic vortex with small but finite $\mathcal{R}$ is shown in Fig. 4.13. Over a time interval long enough for the β effects to become evident, the vortex largely retains its axisymmetric shape, but weakens somewhat, while propagating to the west-southwest as it emits a train of weak-amplitude Rossby waves mostly in its wake. Because of the parity symmetry in the quasigeostrophic shallow-water equations, a cyclonic initial vortex behaves analogously, except that its propagation direction is west-northwest.

One way to understand the vortex propagation induced by β is to recognize that the associated forcing term in (4.114) induces a dipole structure to develop in $\psi(x, y)$ in a situation with a primarily axisymmetric vortical flow, $\psi \approx \Psi(r)$. This is shown by

$$\beta \frac{\partial \psi}{\partial x} \approx \beta \frac{\partial}{\partial x} \Psi(r) = \beta \frac{x}{r} \frac{d\Psi}{dr} = \beta \cos[\theta] \frac{d\Psi}{dr}.$$

The factor $\cos[\theta]$ represents a dipole circulation in ψ. A dipole vortex is an effective advective configuration for spatial propagation (cf., the point-vortex dipole

(a)

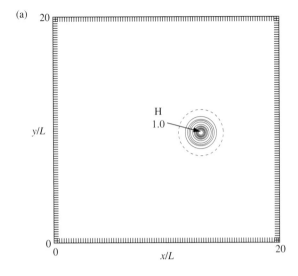

(b)

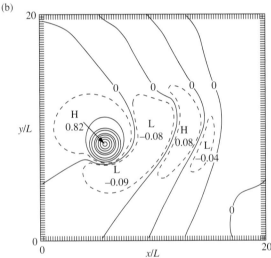

Fig. 4.13. Propagation of a strong anticyclonic vortex on the β-plane: (a) $\psi(x, y)$ at $t = 0$ and (b) $\psi(x, y)$ at $t = 17.3 \times 1/L\beta$. ψ and **x** values are made non-dimensional with the initial vortex amplitude and size scales, VL and L, respectively, and the deformation radius is slightly smaller than the vortex, $R = 0.7L$. The vortex propagates to the west-southwest, approximately intact, while radiating Rossby waves in its wake. (McWilliams & Flierl, 1979)

solution in Section 3.2.1). With further evolution the early-time zonal separation between the dipole centers is rotated by the azimuthal advection associated with Ψ to a more persistently meridional separation between the centers, and the resulting advective effect on both itself and the primary vortex component, Ψ, is approximately westward. The dipole orientation is not one with a precisely meridional separation, so the vortex propagation is not precisely westward.

As azimuthal asymmetries develop in the solution, the advective influence by Ψ acts to suppress them by the axisymmetrization process discussed in Sections 3.4 and 3.5. In the absence of β, the axisymmetrization process would win, and the associated vortex self-propagation mechanism would be suppressed. In the presence of β, there is continual regeneration of the asymmetric component in ψ. Some of this asymmetry in ψ propagates ("leaks") away from the region with vortex recirculation, and in the far-field it satisfies the weak amplitude assumption for Rossby-wave dynamics. However, the leakage rate is much less than it would be without the opposing advective axisymmetrization effect, so the external Rossby-wave field after a comparable evolution period of $\mathcal{O}(1/\beta L)$ is much weaker when $\mathcal{R}$ is small than when $\mathcal{R}$ is large. This efficiency in preserving the vortex pattern even as it propagates is reminiscent of gravity solitary waves or even solitons. The latter are nonlinear wave solutions for non-shallow-water dynamical systems that propagate without change of shape due to a balance of opposing tendencies between (weak) spreading by wave dispersion and (weak) wave steepening (Section 4.4). However, the specific spreading and steepening mechanisms are different for a β-plane vortex than for gravity waves, because here the dominant advective flow direction is perpendicular to the wave propagation and dispersion directions.

By multiplying (4.114) by $\mathbf{x}$ and integrating over the domain, the following equation can be derived for the *centroid* motion:

$$\frac{d}{dt}\mathbf{X} = -\beta R^2, \tag{4.125}$$

where $\mathbf{X}$ is the ψ-weighted centroid for the flow,

$$\mathbf{X} = \frac{\iint \mathbf{x}\psi\,d\mathbf{x}}{\iint \psi\,d\mathbf{x}}. \tag{4.126}$$

(The alternatively defined, point-vortex centroids in Section 3.2.1 have a vertical vorticity, ζ, weighting.) Thus, the centroid propagates westward at the speed of a long Rossby wave. To the extent that a given flow evolution approximately preserves its pattern, as is true for the vortex solution in Fig. 4.13, then the pattern as a whole must move westward with speed, βR^2. This is approximately what happens with the vortex. The emitted Rossby-wave wake also enters into determining $\mathbf{X}$, so it is consistent for the vortex motion to depart somewhat from the centroid motion. For example, the calculated southward vortex motion requires a meridional asymmetry in the Rossby-wave wake, such that the positive ψ extrema are preferentially found northward of both the negative extrema and the primary vortex itself. Note that as a barotropic limit is approached ($R \gg L$), the centroid speed (4.125) increases. Numerical vortex solutions indicate that the Rossby-wave emission rate increases with R/L; the vortex pattern persistence over a $(\beta L)^{-1}$ time scale decreases (i.e., Rossby-wave disperison diminishes); and the vortex propagation rate drops well below the centroid movement rate. In a strictly barotropic limit ($R = \infty$), however, the relation (4.125) cannot be derived from the potential vorticity equation and thus is irrelevant.

The ratio, $\mathcal{R}$, approximately characterizes the boundary between wave propagation and turbulence for barotropic dynamics. The length scale, L_β, that makes $\mathcal{R} = 1$ for a given level of kinetic energy, $\sim V^2$, is defined by

$$L_\beta = \sqrt{\frac{V}{\beta}}. \tag{4.127}$$

The dynamics for flows with a scale of $L > L_\beta$ is essentially a Rossby-wave propagation with a weak advective influence, whereas for $L < L_\beta$, it is 2D turbulence (Section 3.7) with weak β effects (or isolated vortex propagation as in Fig. 4.13). L_β is sometimes called the *Rhines scale* in this context. L_β also is a relevant scale for the width of the western boundary current in an oceanic wind gyre (Section 6.2).

4.8.2 Eastern boundary Kelvin wave

Another mechanism for Rossby-wave emission is the poleward propagation of a Kelvin wave along an eastern boundary (Fig. 4.14). The Kelvin-wave solution in Section 4.2.3 is not valid on the β plane, even though it is reasonable to expect that it will remain approximately valid for waves with a scale smaller than Earth's radius (i.e., $\beta L/f \sim L/a \ll 1$). In particular, as a Kelvin wave moves poleward, the local value for $R = \sqrt{gH}/f$ decreases since $f(y)$ increases (in the absence of a compensating change in H). Since the off-shore scale for the Kelvin wave is R, the spatial structure must somehow adjust to its changing environment, $R(y)$, at a rate of $\mathcal{O}(\beta L)$, if it is to remain approximately a Kelvin wave. While it is certainly an a-priori possibility that the evolution does not remain close to local Kelvin-wave behavior, both theoretical solutions and oceanic observations indicate that it often does so. In a linearized wave analysis for this β-adjustment process, any energy lost to the transmitted Kelvin wave must be scattered into either geostrophic currents, geostrophically balanced Rossby waves, or unbalanced inertia-gravity waves. Because the cross-shore momentum balance for a Kelvin wave is geostrophic, it is perhaps not surprising that most of the scattered energy goes into a coastally trapped, along-shore, geostrophic current, left behind after the passage of the Kelvin wave, and into westward propagating Rossby waves that move into the domain interior (Fig. 4.14). In the El Niño scenario

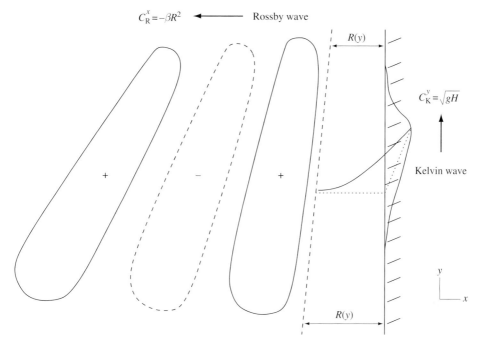

Fig. 4.14. Rossby-wave emission from a poleward Kelvin wave along an eastern boundary on the β-plane. The meridional and zonal propagation velocities are C_K^y and C_R^x, respectively.

in Fig. 4.7, a deepening of the eastern equatorial pycnocline, in association with a surface temperature warming, instigates a poleward-propagating Kelvin wave that lowers the pycnocline all along the eastern boundary. This implies a thermal-wind balanced, along-shore flow close to the coast with poleward vertical shear. With the interpretation of this shallow-water solution as equivalent to first baroclinic vertical mode (Section 5.1), the along-shore flow has a vertical structure of a poleward surface current and an equatorward undercurrent. (Currents with this structure, as well as with its opposite sign, are frequently observed along sub-tropical eastern boundaries.) $\beta \neq 0$ causes both the passing Kelvin wave and its along-shore flow wake to emit long, reduced-gravity (i.e., baroclinic) Rossby waves westward into the oceanic interior. Since $R(y)$, hence $\mathcal{B}(y)$ decreases away from the Equator, so do the zonal phase and group velocities in the long wave limit (4.123); as a consequence the emitted Rossby-wave crests bend outward from the coastline closer to the Equator (Fig. 4.14).

The analogous situation for Kelvin waves propagating equatorward along a western boundary does not have an efficient Rossby-wave emission process. Kelvin waves have an off-shore, zonal scale $\sim R$, and their scattering is most efficient into motions with a zonal similar scale. But the zonal group velocity for

Rossby waves (4.122) implies they cannot propagate energy eastward ($c_{\mathrm{g}}^x > 0$) unless their cross-shore scale is much smaller than R. However, for a given wave amplitude (e.g., V or ψ_0 in (4.119)), the nonlinearity measure, $\mathcal{R}^{-1}$ from (4.124), increases as L decreases, so emitted Rossby waves near a western boundary are much more likely to evolve in a turbulent rather than wave-like manner. Furthermore, a western boundary region is usually occupied by strong currents due to wind gyres (Chapter 6), and this further adds to the advective dynamics of any emitted Rossby waves. The net effect is that most of the Kelvin-wave scattering near a western boundary goes into along-shore geostrophic currents that remain near the boundary rather than Rossby waves departing from the boundary region.

5

Baroclinic and jet dynamics

The principal mean circulation patterns for the ocean and atmosphere are unstable to perturbations. Therefore, the general circulation is intrinsically variable, even with periodic solar forcing, invariant oceanic and atmospheric chemical compositions, and fixed land and sea-floor topography – none of which is absolutely true at any time scale nor even approximately true over millions of years. The statistics of its variability may be considered stationary in time under these steady-state external influences; i.e., it has an unsteady *statistical equilibrium* dynamics comprised of externally forced, but unstable, mean flows and turbulent eddies, waves, and vortices that are generated by the instabilities. In turn, the mean–eddy fluxes of momentum, heat, potential vorticity, and material tracers reshape the structure of the mean circulation and material distributions, evinced by their importance in mean dynamical balance equations (e.g., Section 3.4).

Where the mean circulations have a large spatial scale and are approximately geostrophic and hydrostatic, the important instabilities are also geostrophic, hydrostatic, and somewhat large scale (i.e., synoptic scale or mesoscale). These instabilities are broadly grouped into two classes.

- **Barotropic instability:** the mean horizontal shear is the principal energy source for the eddies, and horizontal momentum flux (Reynolds stress) is the dominant eddy flux (Chapter 3).
- **Baroclinic instability:** the mean vertical shear and horizontal buoyancy gradient (related through the thermal wind) is the energy source, and vertical momentum and horizontal buoyancy fluxes are the dominant eddy fluxes, with Reynolds stress playing a secondary role.

Under some circumstances the mean flows are unstable to other, smaller-scale types of instability (e.g., convective, Kelvin–Helmholtz, or centrifugal), but these are relatively rare as direct instabilities of the mean flows on the planetary scale. More often these instabilities arise either in response to locally forced flows (e.g., in boundary layers; Chapter 6) or as secondary instabilities of synoptic and mesoscale flows as part of a general cascade of variance toward dissipation on very small scales.

The mean zonal wind in the troposphere (Fig. 5.1) is a geostrophic flow with an associated meridional temperature gradient created by tropical heating and polar cooling. This wind profile is baroclinically unstable to extra-tropical fluctuations on the synoptic scale of $\mathcal{O}(10^3)$ km. This is the primary origin of weather, and in turn weather events collectively cause a poleward heat flux that limits the strength of the zonal wind and its geostrophically balancing meridional temperature gradient.

In this chapter baroclinic instability is analyzed in its simplest configuration as a two-layer flow. To illustrate the finite-amplitude, long-time consequences (i.e., the eddy–mean interaction for a baroclinic flow; cf. Section 3.4), an idealized problem for the statistical equilibrium of a baroclinic zonal jet is analyzed at the end of the chapter. Of course, there are many other aspects of baroclinic dynamics (e.g., vortices and waves) that are analogous to their barotropic and shallow-water counterparts, but these topics will not be revisited in this chapter.

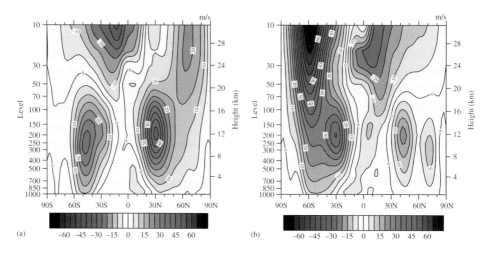

Fig. 5.1. Color plate 4.   The mean zonal wind [m s^{-1}] for the atmosphere, averaged over time and longitude during 2003: (left) January and (right) July. The vertical axis is labeled both by height [km] and by pressure level [1 mb $= 10^2$ Pa]. Note the following features: the eastward velocity maxima at the tropopause (at $\approx 200 \times 10^2$ Pa) that are stronger in the winter hemisphere; the wintertime stratospheric polar night jet; the stratospheric tropical easterlies that shift hemisphere with the seasons; and the weak westward surface winds both in the tropics (i.e., trade winds) and near the poles. (National Centers for Environmental Prediction climatological analysis (Kalnay *et al.*, 1996), courtesy of Dennis Shea, National Center for Atmospheric Research.)

5.1 Layered hydrostatic model

5.1.1 Two-layer equations

Consider the governing equations for two immiscible (i.e., unmixing) fluid layers, each with constant density and with the upper layer ($n = 1$) fluid lighter than the lower layer ($n = 2$) fluid, as required for gravitational stability (Fig. 5.2). When the fluid motions are sufficiently thin ($H/L \ll 1$), hence hydrostatic, each of the layers has a shallow-water dynamics, except they are also dynamically coupled through the pressure-gradient force. To derive this coupling, make an integration of the hydrostatic balance relation downward from the rigid top surface at $z = H$ (i.e., as in Section 4.1):

$$p = p_{\mathrm{H}}(x, y, t) \text{ at } z = H$$

$$p_1 = p_{\mathrm{H}} + \rho_1 g(H - z) \text{ in layer 1}$$

$$p_2 = p_{\mathrm{H}} + \rho_1 g(H - h_2) + \rho_2 g(h_2 - z) \text{ in layer 2.} \tag{5.1}$$

Thus,

$$\nabla p_1 = \nabla \, p_{\mathrm{H}}$$

$$\nabla p_2 = \nabla \, p_{\mathrm{H}} + g_1' \rho_0 \nabla h_2, \tag{5.2}$$

and

$$g_1' = g \frac{\rho_2 - \rho_1}{\rho_0} > 0 \tag{5.3}$$

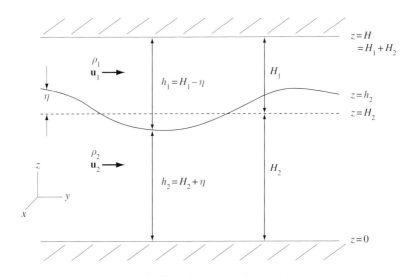

Fig. 5.2. Sketch of a two-layer fluid.

is the *reduced gravity* associated with the relative density difference across the interface between the layers (cf. (4.11)). Here $\mathbf{V} = \mathbf{V}_h$ is the horizontal gradient operator. Expressed in terms of the interface displacement relative to its resting position, η, and layer geopotential functions, $\phi_n = p_n/\rho_0$ for $n = 1, 2$, (5.2) and (5.3) imply that

$$\eta = -\frac{\phi_1 - \phi_2}{g_1'}, \tag{5.4}$$

and the layer thicknesses are

$$h_1 = H_1 - \eta, \quad h_2 = H_2 + \eta, \quad h_1 + h_2 = H_1 + H_2 = H. \tag{5.5}$$

In each layer the Boussinesq horizontal momentum and mass balances are

$$\frac{D\mathbf{u}_n}{Dt_n} + f\hat{\mathbf{z}} \times \mathbf{u}_n = -\mathbf{V}\phi_n + \mathbf{F}_n$$

$$\frac{\partial h_n}{\partial t} + \mathbf{V} \cdot (h_n \mathbf{u}_n) = 0, \tag{5.6}$$

for $n = 1, 2$. The substantial derivative in each layer is

$$\frac{D}{Dt_n} = \frac{\partial}{\partial t} + \mathbf{u}_n \cdot \mathbf{V}.$$

It contains only horizontal advection as a result of the assumption that both the horizontal velocity, $\mathbf{u}_n$, and the advected quantity are depth-independent within each layer (as in the shallow-water equations; Section 4.1). This partial differential equation system is the primitive equations in uniform-density layers.

After applying the curl operator to the momentum equation in (5.6), the resulting horizontal divergence, $\mathbf{V} \cdot \mathbf{u}$, can be eliminated using the thickness equation. The result in each layer is the two-layer potential vorticity equation,

$$\frac{Dq_n}{Dt_n} = \mathcal{F}_n,$$

$$q_n = \frac{f(y) + \zeta}{h_n},$$

$$\zeta_n = \hat{\mathbf{z}} \cdot \mathbf{V} \times \mathbf{u}_n,$$

$$\mathcal{F}_n = \hat{\mathbf{z}} \cdot \mathbf{V} \times \mathbf{F}_n, \tag{5.7}$$

with layer potential vorticity, q_n; relative vorticity, ζ_n; and force curl, $\mathcal{F}_n$. This q definition and its governing equation are essentially similar to the shallow-water potential vorticity relations (Section 4.1.1), except that here they hold true for each separate layer.

The vertical velocity at the interface is determined by the kinematic condition (Section 2.2.3),

$$w_{\mathrm{I}} = \frac{D\eta}{Dt} = \frac{Dh_2}{Dt} = \frac{D}{Dt}(H - h_1) = -\frac{Dh_1}{Dt}. \tag{5.8}$$

There is an ambiguity about which advecting velocity to use in (5.8) since $\mathbf{u}_1 \neq \mathbf{u}_2$. These two choices give different values for w_{I}. Since $w = 0$ at the boundaries $(z = 0, H)$ in the absence of boundary stress (cf., Section 5.3), the vertical velocity is determined at all heights as a piecewise linear function that connects the boundary and interfacial values within each layer, but with a discontinuity in the value of w at the interface. So a disconcerting feature for a layered hydrostatic model is that the 3D velocities, as well as the layer densities by the definition of the model, are discontinuous at the interface, although the pressure (5.1) is continuous.

The background density profile is given by the ρ_n in a resting state configuration with $\mathbf{u}_n = \eta = w_{\mathrm{I}} = 0$. Because the density is constant within each layer, there are no density changes following a fluid parcel since the parcels remain within layers. Nevertheless, an auxiliary interpretation of the moving interface is that it induces a density fluctuation, ρ'_{I}, or equivalently a buoyancy fluctuation, $b_{\mathrm{I}} = -g\rho'_{\mathrm{I}}/\rho_0$, in the vicinity of the interface due to the distortion of the background density profile by $\eta \neq 0$:

$$b_{\mathrm{I}} \approx \frac{g}{\rho_0}\frac{d\overline{\rho}}{dz}\Big|_{\mathrm{I}}\eta = -\frac{2g'_{\mathrm{I}}}{H}\eta = \frac{2}{H}(\phi_1 - \phi_2). \tag{5.9}$$

The last relation expresses hydrostatic balance across the layer interface.

Now make a quasigeostrophic approximation for the two-layer equations, analogous to that for the shallow-water equations (Section 4.6).

$$\mathbf{u}_{g,n} = \hat{\mathbf{z}} \times \nabla\psi_n, \quad \psi_n = \frac{1}{f_0}\phi_n,$$

$$\frac{D}{Dt_{g,n}}[\zeta_n + \beta y] = f_0\frac{\partial w_n}{\partial z} + \mathcal{F}_n,$$

$$\frac{D}{Dt_{g,n}} = \frac{\partial}{\partial t} + \mathbf{u}_{g,n}\cdot\nabla. \tag{5.10}$$

$\mathbf{u}_{g,n}$ is the geostrophic velocity in layer n, and $D_{t_{g,n}}$ is its associated substantial derivative.

Note that

$$\delta h_2 = -\frac{\delta p_1 - \delta p_2}{g\Delta\rho} = -\delta h_1$$

from (5.1). Hence, using the geostrophic approximation and the linear dependence of w within the layers,

$$\frac{\partial w_1}{\partial z} = \frac{w(H) - w_1}{h_1}$$

$$\approx -\frac{w_1}{H_1}$$

$$\approx \frac{f_0}{g_1' H_1} \frac{D}{Dt_{g,1}} (\psi_1 - \psi_2). \tag{5.11}$$

Similarly,

$$\frac{\partial w_2}{\partial z} \approx -\frac{f_0}{g_1' H_2} \frac{D}{Dt_{g,2}} (\psi_1 - \psi_2). \tag{5.12}$$

There is no discontinuity associated with which layer's D_t appears in (5.11) and (5.12) since

$$\frac{D(\psi_1 - \psi_2)}{Dt_{g,1}} = \frac{D(\psi_1 - \psi_2)}{Dt_{g,2}}$$

using the quasigeostrophic approximations. As with the shallow-water equations, these two-layer relations can be combined into the quasigeostrophic potential vorticity equations for a two-layer fluid (analogous to the primitive-equation form in (5.7)):

$$\frac{Dq_{QG,n}}{Dt_{g,n}} = \mathcal{F}_n, \tag{5.13}$$

with the potential vorticities and substantial derivatives defined by

$$q_{QG,1} = \nabla^2 \psi_1 + \beta y - \frac{f_0^2}{g_1' H_1} (\psi_1 - \psi_2)$$

$$q_{QG,2} = \nabla^2 \psi_2 + \beta y + \frac{f_0^2}{g_1' H_2} (\psi_1 - \psi_2)$$

$$\frac{D}{Dt_{g,n}} = \partial_t + J[\psi_n,]. \tag{5.14}$$

The energy conservation principle for a two-layer model is a straightforward generalization of the rotating shallow-water model, and it is derived by a similar path (Section 4.1.1). The relation for the primitive equations is

$$\frac{dE}{dt} = 0, \tag{5.15}$$

where

$$E = \iint dx\,dy\,\frac{1}{2}\left(h_1 u_1^2 + h_2 u_2^2 + g_1' \eta^2\right) \tag{5.16}$$

(cf. (4.17)). Its integrand is comprised of depth-integrated horizontal kinetic energy and available potential energy. (There is no internal energy component in E here because compressional work has been neglected in this Boussinesq- and primitive-equation model; Section 2.1.4.) In the quasigeostrophic approximation, the principle (5.15) applies to the simpler energy,

$$E \equiv \iint dx\,dy\,\frac{1}{2}\left(H_1(\nabla\psi_1)^2 + H_2(\nabla\psi_2)^2\right.$$
$$\left. + \frac{f^2}{g_1'}(\psi_1 - \psi_2)^2\right) \tag{5.17}$$

(cf., (4.115)).

5.1.2 N-layer equations

From the preceding derivation it is easy to imagine the generalization to N layers with a monotonically increasing density profile with depth, $\rho_{n+1} > \rho_n \,\forall\, n \leq N-1$, while continuing to make the hydrostatic assumption and assume that the layers neither mix nor overturn. The result can be expressed as the following horizontal momentum and mass balances:

$$\frac{D\mathbf{u}_n}{Dt_n} + f\hat{\mathbf{z}} \times \mathbf{u}_n = -\nabla\phi_n + \mathbf{F}_n$$

$$\frac{\partial h_n}{\partial t} + \nabla \cdot (h_n \mathbf{u}_n) = 0 \tag{5.18}$$

for $n = 1, \ldots, N$. This is isomorphic to (5.6) except that there is an expanded range for n. Thus, the N-layer potential vorticity equation is also isomorphic to (5.7). Accompanying these equations are several auxiliary relations. The layer thicknesses are

$$h_1 = H_1 - \eta_{1.5}$$
$$h_n = H_n + \eta_{n-0.5} - \eta_{n+0.5}, \quad 2 \leq n \leq N-1$$
$$h_N = H_N + \eta_{N-0.5}. \tag{5.19}$$

H_n is the resting layer depth, and $\eta_{n+0.5}$ is the interfacial displacement between layers n and $n+1$. The hydrostatic geopotential function is

$$g_{n+0.5}' \eta_{n+0.5} = \phi_{n+1} - \phi_n, \quad n = 1, \ldots, N-1, \tag{5.20}$$

and

$$g'_{n+0.5} = g \frac{p_{n+1} - p_n}{p_0} \tag{5.21}$$

is the reduced gravity for the interface $n + 0.5$. The vertical velocity at the interfaces is

$$w_{n+0.5} = \frac{D\eta_{n+0.5}}{Dt}, \qquad n = 1, \ldots, N - 1, \tag{5.22}$$

and the buoyancy field is

$$b_{n+0.5} = -\frac{2g'_{n+0.5}}{H_n + H_{n+1}} \eta_{n+0.5}, \qquad n = 1, \ldots, N - 1. \tag{5.23}$$

Because of the evident similarity among the governing equations, this N-layer model is often called the *stacked shallow-water model*. It represents a particular vertical discretization of the adiabatic primitive equations expressed in a trans-formed *isentropic coordinate system* (i.e., (x, y, ρ, t), analogous to the pressure coordinates, $(x, y, F(p), t)$, in Section 2.3.5). As $N \to \infty$, (5.18)–(5.23) converge to the continuously stratified, 3D primitive equations in isentropic coordinates for solutions that are sufficiently vertically smooth. An interpretive attractiveness of the *isentropic primitive equations* is the disappearance of any explicit vertical velocity in the advection operator; this occurs as a result of the parcel conservation of (potential) density. Another interpretive advantage is the relative simplicity of the definition for the potential vorticity for the isentropic primitive equations (another parcel invariant for conservative dynamics), namely,

$$q_{\text{IPE}} = \frac{f + \zeta}{\partial Z / \partial \rho}, \tag{5.24}$$

with $Z(x, y, \rho, t)$ the height of an isentropic surface and ζ determined from horizontal derivatives of $\mathbf{u}_h$ at constant ρ, i.e., within the isentropic coordinates. Equation (5.24) is to be compared to its equivalent but more complicated expression in Cartesian coordinates, known as *Ertel potential vorticity* in a hydrostatic approximation,

$$q_E = (f + \zeta) \frac{\partial \rho}{\partial z} + \frac{\partial u}{\partial z} \frac{\partial \rho}{\partial y} - \frac{\partial v}{\partial z} \frac{\partial \rho}{\partial x} \tag{5.25}$$

with ζ determined from horizontal derivatives of $\mathbf{u}_h$ at constant z.

Similarly there is a straightforward generalization for the quasigeostrophic potential vorticity equations and streamfunction, ψ_n, in an N-layer model:

$$\frac{Dq_{\text{QG},n}}{Dt_{g,n}} = \mathcal{F}_n. \tag{5.26}$$

The layer potential vorticities are defined by

$$q_{QG,1} = \nabla^2 \psi_1 + \beta y - \frac{f_0^2}{g_{1.5}' H_1}(\psi_1 - \psi_2),$$

$$q_{QG,n} = \nabla^2 \psi_n + \beta y - \frac{f_0^2}{g_{n+0.5}' H_n}(\psi_n - \psi_{n+1}) + \frac{f_0^2}{g_{n-0.5}' H_n}(\psi_{n-1} - \psi_n), \ 1 < n < N$$

$$q_{QG,N} = \nabla^2 \psi_N + \beta y + \frac{f_0^2}{g_{N-0.5}' H_N}(\psi_{N-1} - \psi_N), \tag{5.27}$$

and the quasigeostrophic substantial derivative is

$$\frac{D}{Dt_{g,n}} = \partial_t + J[\psi_n, \ ].$$

The layer potential vorticity is a vertical-layer discretization for the vertically continuous, 3D quasigeostrophic potential vorticity defined by

$$q_{QG}(x, y, z, t) = \nabla^2 \psi + \beta y + \frac{\partial}{\partial z}\left(\frac{f_0^2}{\mathcal{N}^2}\frac{\partial \psi}{\partial z}\right). \tag{5.28}$$

$\mathcal{N}(z)$ is the continuous buoyancy or Brünt–Väisällä frequency defined in (2.70) (NB, a different symbol is used here to avoid confusion with the discrete layer number, N). $\mathcal{N}$ is the limiting form for $\sqrt{g'/\Delta H}$ as the layer thickness, ΔH, becomes vanishingly small. The finite-difference operations among the ψ_n in (5.27) are discrete approximations to the continuous vertical derivatives in (5.28). (The expression (5.28) is derived in Exercise 11 of the preceding chapter.)

5.1.3 Vertical modes

In a baroclinic fluid – whether in an N-layer or a continuously stratified model and whether in the full Boussinesq or approximate primitive or quasigeostrophic dynamics – it is common practice to decompose the fields into *vertical modes*. This is analogous to making a Fourier transform with respect to the vertical height or density coordinate. Formally each dependent variable (e.g., streamfunction, $\psi(x, y, z, t)$) can be written as a discrete summation over the vertical modal contributions (e.g., $\tilde{\psi}(x, y, m, t)$). The vertical modes, G_m, are discrete functions in a discrete layered model,

$$\psi_n(x, y, t) = \sum_{m=0}^{N-1} \tilde{\psi}_m(x, y, t) \, G_m(n), \tag{5.29}$$

and they are continuous functions in continuous height coordinates,

$$\psi(x, y, z, t) = \sum_{m=0}^{\infty} \tilde{\psi}_m(x, y, t) G_m(z). \tag{5.30}$$

Each G_m is determined in order to "diagonalize" the dynamical coupling among different vertical layers or levels for the linear terms in the governing equations above, e.g., in (5.26) and (5.27) for quasigeostrophic dynamics. The transformation converts the coupling among layer variables to uncoupled modal variables by the technique described in the next paragraph. The modal transformation is analogous to using a Fourier transform to convert a spatial derivative, $\mathbf{V}$, that couples $\psi(\mathbf{x})$ in neighboring locations, into a simple algebraic factor, $i\mathbf{k}$, that only multiplies the local modal amplitude, $\tilde{\psi}(\mathbf{k})$ (Section 3.7). The vertical scale of the G_m decreases with increasing m, as with a Fourier transform where the spatial scale decreases with increasing $|\mathbf{k}|$. The prescription for determining G_m given below also makes the modes orthonormal, e.g., for a layered model,

$$\sum_{n=1}^{N} \frac{H_n}{H} G_p(n) \, G_q(n) = \delta_{p,q}, \tag{5.31}$$

or for continuous height modes,

$$\frac{1}{H} \int_0^H dz \, G_p(z) \, G_q(z) = \delta_{p,q}, \tag{5.32}$$

with $\delta_{p,q} = 1$ if $p = q$, and $\delta_{p,q} = 0$ if $p \neq q$ (i.e., δ is a discrete delta function). This is a mathematically desirable property for a set of vertical *basis functions* because it assures that the inverse transformation for (5.30) is well defined as

$$\tilde{\psi}_m = \sum_{n=1}^{N} \frac{H_n}{H} \psi_n \, G_m(n) \tag{5.33}$$

or

$$\tilde{\psi}_m = \frac{1}{H} \int_0^H dz \, \psi(z) G_m(z). \tag{5.34}$$

The physical motivation for making this transformation comes from measurements of large-scale atmospheric and oceanic flows that show that most of the energy is associated with only a few of the gravest vertical modes (i.e., those with the smallest m values and correspondingly largest vertical scales). So it is more efficient to analyze the behavior of $\tilde{\psi}_m(x, y, t)$ for a few m values than of $\psi(x, y, z, t)$ at all z values with significant energy. A more theoretical motivation is that the vertical modes can be chosen – as explained in the rest of this section – so that each mode has an independent (i.e., decoupled from other modes) linear dynamics analogous to a single fluid layer (barotropic or shallow-water). In general, a full dynamical decoupling between the vertical modes cannot be achieved, but it can be done for some important behaviors, e.g., the Rossby-wave propagation in Section 5.2.1.

For specificity, consider the two-layer quasigeostrophic equations ($N = 2$) to illustrate how the G_m are calculated. The two vertical modes are referred to as barotropic ($m = 0$) and baroclinic ($m = 1$). (For an N-layer model, each mode with $m \geq 1$ is referred to as the mth baroclinic mode.) To achieve the linear-dynamical decoupling between layers, it is sufficient to "diagonalize" the relationship between the potential vorticity and streamfunction. That is, determine the 2×2 matrix $G_m(n)$ such that each modal potential vorticity contribution (apart from the planetary vorticity term), i.e.,

$$\tilde{q}_{\mathrm{QG},m} - \beta y = \frac{1}{H} \sum_{n=1}^{2} H_n \left(q_{\mathrm{QG},n} - \beta y \right) G_m(n),$$

depends only on its own modal streamfunction field,

$$\tilde{\psi}_m = \frac{1}{H} \sum_{n=1}^{2} H_n \psi_n G_m(n),$$

and not on any other $\tilde{\psi}_{m'}$ with $m' \neq m$. This is accomplished by the following choice:

$$G_0(1) = 1 \qquad G_0(2) = 1 \quad \text{(barotropic mode)}$$

$$G_1(1) = \sqrt{\frac{H_2}{H_1}} \qquad G_1(2) = -\sqrt{\frac{H_1}{H_2}} \quad \text{(baroclinic mode)} \qquad (5.35)$$

as can be verified by applying the operator $H^{-1} \sum_{n=1}^{2} H_n G_m(n)$ to (5.14) and substituting these G_m values. The barotropic mode is independent of height, while the baroclinic mode reverses its sign with height and has a larger amplitude in the thinner layer. Both modes are normalized as in (5.31).

With this choice for the vertical modes, the modal streamfunction fields are related to the layer streamfunctions by

$$\tilde{\psi}_0 = \frac{H_1}{H} \psi_1 + \frac{H_2}{H} \psi_2$$

$$\tilde{\psi}_1 = \frac{\sqrt{H_1 H_2}}{H} \left(\psi_1 - \psi_2 \right), \qquad (5.36)$$

and the inverse relations for the layer streamfunctions are

$$\psi_1 = \tilde{\psi}_0 + \frac{H_2}{H_1} \tilde{\psi}_1$$

$$\psi_2 = \tilde{\psi}_0 - \frac{H_1}{H_2} \tilde{\psi}_1. \qquad (5.37)$$

The barotropic mode is therefore the depth average of the layer quantities, and the baroclinic mode is proportional to the deviation from the depth average.

The various factors involving H_n assure the orthonormality property (5.32). Identical linear combinations relate the modal and layer potential vorticities, and after substituting from (5.14), the latter are evaluated to be

$$\tilde{q}_{QG,0} = \beta y + \nabla^2 \tilde{\psi}_0$$

$$\tilde{q}_{QG,1} = \beta y + \nabla^2 \tilde{\psi}_1 - \frac{1}{R_1^2} \tilde{\psi}_1. \tag{5.38}$$

These relations exhibit the desired decoupling among the modal streamfunction fields. Here the quantity,

$$R_1^2 = \frac{g' H_1 H_2}{f_0^2 H}, \tag{5.39}$$

defines the deformation radius for the baroclinic mode, R_1. By analogy, since the final term in $\tilde{q}_{QG,1}$ has no counterpart in $\tilde{q}_{QG,0}$, the two modal $\tilde{q}_{QG,m}$ can be said to have an identical definition in terms of $\tilde{\psi}_m$ if the barotropic deformation radius is defined to be

$$R_0 = \infty. \tag{5.40}$$

The form of (5.38) is the same as the quasigeostrophic potential vorticity for barotropic and shallow-water fluids, (3.28) and (4.113), with the corresponding deformation radii, $R = \infty$ and $R = \sqrt{gH}/f_0$, respectively.

This procedure for deriving the vertical modes, G_m, can be expressed in matrix notation for arbitrary N. The layer potential vorticity and streamfunction vectors,

$$\mathbf{q}_{QG} = \{q_{QG,n}; n = 1, \ldots, N\} \quad \text{and} \quad \boldsymbol{\psi} = \{\psi_n; n = 1, \ldots, N\},$$

are related by (5.27) re-expressed as

$$\mathbf{q}_{QG} = P\boldsymbol{\psi} + \mathbf{I}\beta y. \tag{5.41}$$

Here $\mathbf{I}$ is the identity vector (i.e., equal to one for every element), and P is the matrix operator that represents the contribution of $\boldsymbol{\psi}$ derivatives to $\mathbf{q}_{QG} - \mathbf{I}\beta y$, namely,

$$P = I\nabla^2 - S, \tag{5.42}$$

where I is the identity matrix; $I\nabla^2$ is the relative vorticity matrix operator; and S, the stretching vorticity matrix operator, represents the cross-layer coupling. The modal transformations (5.30) and (5.33) are expressed in matrix notation as

$$\boldsymbol{\psi} = G\tilde{\boldsymbol{\psi}}, \quad \tilde{\boldsymbol{\psi}} = G^{-1}\boldsymbol{\psi}, \tag{5.43}$$

with analogous expressions relating $\mathbf{q}_{QG} - \mathbf{I}\beta y$ and $\tilde{\mathbf{q}}_{QG} - \mathbf{I}\beta y$. The matrix G is related to the functions in (5.29) by $G_{nm} = G_m(n)$. Thus,

$$\tilde{\mathbf{q}}_{QG} = G^{-1} PG\tilde{\boldsymbol{\psi}} + \tilde{\mathbf{I}}_0 \beta y = \left[I \nabla^2 - G^{-1} SG \right] \tilde{\boldsymbol{\psi}} + \tilde{\mathbf{I}}\beta y, \tag{5.44}$$

using $G^{-1}G = I$.

Therefore, the goal of eliminating cross-modal coupling in (5.44) is accomplished by making $G^{-1}SG$ a diagonal matrix, i.e., by choosing the vertical modes, $G = G_m(n)$, as eigenmodes of S with corresponding eigenvalues, $R_m^{-2} \geq 0$, such that

$$SG - R^{-2}G = 0 \tag{5.45}$$

for the diagonal matrix, $R^{-2} = \delta_{n,m} R_m^{-2}$. As in (5.39) and (5.40), R_m is called the deformation radius for the mth eigenmode. From (5.27), S is defined by

$$S_{11} = \frac{f_0^2}{g'_{1.5}H_1}, \quad S_{12} = \frac{-f_0^2}{g'_{1.5}H_1}, \quad S_{1n} = 0, n > 2$$

$$S_{21} = \frac{-f_0^2}{g'_{1.5}H_2}, \quad S_{22} = \frac{f_0^2}{H_2}\left(\frac{1}{g'_{1.5}} + \frac{1}{g'_{2.5}}\right), \quad S_{23} = \frac{-f_0^2}{g'_{2.5}H_2}, \quad S_{2n} = 0, n > 3$$

$$\ldots$$

$$S_{Nn} = 0, n < N-1, \quad S_{NN-1} = \frac{-f_0^2}{g'_{N-0.5}H_N}, \quad S_{NN} = \frac{f_0^2}{g'_{N-0.5}H_N}. \tag{5.46}$$

For $N = 2$ in particular,

$$S_{11} = \frac{f_0^2}{g'_I H_1}, \quad S_{12} = \frac{-f_0^2}{g'_I H_1},$$

$$S_{21} = \frac{-f_0^2}{g'_I H_2}, \quad S_{22} = \frac{f_0^2}{g'_I H_2}. \tag{5.47}$$

It can readily be shown that (5.35), (5.39), and (5.40) are the correct eigenmodes and eigenvalues for this S matrix.

S can be recognized as the negative of a layer-discretized form of a second vertical derivative with unequal layer thicknesses. Thus, just as (5.28) is the continuous limit for the discrete layer potential vorticity in (5.27), the continuous limit for the vertical modal problem (5.45) is

$$\frac{d}{dz}\left(\frac{f_0^2}{\mathcal{N}^2}\frac{dG}{dz}\right) + R^{-2}G = 0. \tag{5.48}$$

Vertical boundary conditions are required to make this a well-posed boundary-eigenvalue problem for $G_m(z)$ and R_m. From (5.20)–(5.22) the vertically continuous formula for the quasigeostrophic vertical velocity is

$$w_{\mathrm{QG}} = \frac{f_0}{\mathcal{N}^2}\frac{D}{Dt}_g\left(\frac{\partial\psi}{\partial z}\right). \tag{5.49}$$

Zero vertical velocity at the boundaries is assured by $\partial\psi/\partial z = 0$, so an appropriate boundary condition for (5.48) is

$$\frac{dG}{dz} = 0 \quad \text{at} \quad z = 0, H. \tag{5.50}$$

When $\mathcal{N}^2(z) > 0$ at all heights, the eigenvalues from (5.48) and (5.50) are countably infinite in number, positive in sign, and ordered by magnitude: $R_0 > R_1 > R_2 > \cdots > 0$. The eigenmodes satisfy the orthonormality condition (5.32). Figure 5.3 illustrates the shapes of the $G_m(z)$ for the first few m with a stratification profile, $\mathcal{N}(z)$, that is upward-intensified. For $m = 0$ (barotropic mode), $G_0(z) = 1$, corresponding to $R_0 = \infty$. For $m \geq 1$ (baroclinic modes), $G_m(z)$ has precisely m zero-crossings in z, so larger m corresponds to smaller vertical scales and smaller deformation radii, R_m. Note that the discrete modes in (5.35) for $N = 2$ have the same structure as in Fig. 5.3, except for having a finite truncation level, $M = N - 1$. (The relation, $H_1 > H_2$, in (5.35) is analogous to an upward-intensified $\mathcal{N}(z)$ profile.)

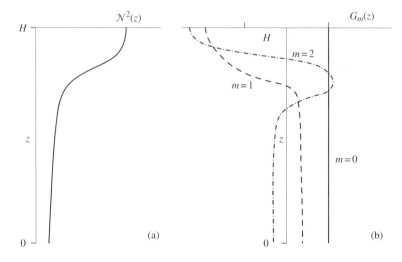

Fig. 5.3. Dynamically determined vertical modes for a continuously stratified fluid: (a) stratification profile, $\mathcal{N}^2(z)$; (b) vertical modes, $G_m(z)$ for $m = 0, 1, 2$.

5.2 Baroclinic instability

The two-layer quasigeostrophic model is now used to examine the stability problem for a mean zonal current with vertical shear (Fig. 5.4). This is the simplest flow configuration exhibiting baroclinic instability (cf., the 3D baroclinic instability in Exercise 8 of this chapter). Even though the shallow-water equations (Chapter 4) contain some of the combined effects of rotation and stratification, they do so incompletely compared to fully 3D dynamics and, in particular, do not admit baroclinic instability because they cannot represent vertical shear.

In this analysis, for simplicity, assume that $H_1 = H_2 = H/2$; hence the baroclinic deformation radius (5.39) is

$$R = \sqrt{g_1' H} \frac{1}{2f}.$$

This choice is a conventional idealization for the stratification in the mid-latitude troposphere, whose mean stability profile, $\mathcal{N}(z)$, is approximately constant in z above the planetary boundary layer (Chapter 6) and below the tropopause. Further assume that there is no horizontal shear (thereby precluding any barotropic instability) and no barotropic component to the mean flow:

$$\overline{\mathbf{u}}_n = (-1)^{(n+1)} U \hat{\mathbf{x}}, \tag{5.51}$$

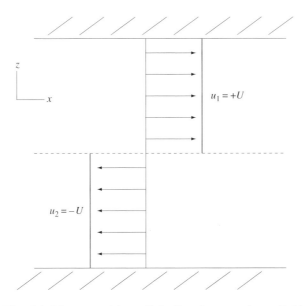

Fig. 5.4. Mean zonal baroclinic flow in a two-layer fluid.

with U a constant. Geostrophically and hydrostatically related mean fields are

$$\overline{\psi}_n = (-1)^{n+1} U y$$

$$\overline{h}_2 = -\frac{f_0}{g'}(\overline{\psi}_1 - \overline{\psi}_2) + \frac{H}{2} = \frac{2 f_0 U y}{g'_{\mathrm{I}}} + \frac{H}{2}$$

$$\overline{h}_1 = H - \overline{h}_2$$

$$\overline{q}_{\mathrm{QG},n} = \beta y + (-1)^{n+1} \frac{U y}{R^2}. \tag{5.52}$$

In this configuration there is more light fluid to the south (in the northern hemi-sphere), since $\overline{h}_2 - H_2 < 0$ for $y < 0$, and more heavy fluid to the north. Making an association between light density and warm temperature, then the south is also warmer and more buoyant (cf., (5.9)). This is similar to the mid-latitude, northern-hemisphere atmosphere, with stronger westerly winds aloft (Fig. 5.1) and warmer air to the south.

Note that (5.51) and (5.52) is a conservative stationary state; i.e., $\partial_t = 0$ in (5.7) if $\mathscr{F}_n = 0$. The $\overline{q}_{\mathrm{QG},n}$ are functions only of y, as are the $\overline{\psi}_n$. So they are functionals of each other. Therefore, $J[\overline{\psi}_n, \overline{q}_{\mathrm{QG},n}] = 0$, and $\partial_t \overline{q}_{\mathrm{QG},n} = 0$. The fluctuation dynamics are linearized around this stationary state. Define

$$\psi_n = \overline{\psi}_n + \psi'_n$$

$$q_{\mathrm{QG},n} = \overline{q}_{\mathrm{QG},n} + q'_{\mathrm{QG},n}, \tag{5.53}$$

and insert these into (5.13) and (5.14), neglecting purely mean terms, perturbation nonlinear terms (assuming weak perturbations), and non-conservative terms:

$$\frac{\partial q'_{\mathrm{QG},n}}{\partial t} + \overline{u}_n \frac{\partial q'_{\mathrm{QG},n}}{\partial x} + v'_n \frac{\partial \overline{q}_{\mathrm{QG},n}}{\partial y} = 0, \tag{5.54}$$

or, evaluating the mean quantities explicitly,

$$\frac{\partial q'_{\mathrm{QG},1}}{\partial t} + U \frac{\partial q'_{\mathrm{QG},1}}{\partial x} + v'_1 \left[\beta + \frac{U}{R^2} \right] = 0$$

$$\frac{\partial q'_{\mathrm{QG},2}}{\partial t} - U \frac{\partial q'_{\mathrm{QG},2}}{\partial x} + v'_2 \left[\beta - \frac{U}{R^2} \right] = 0. \tag{5.55}$$

5.2.1 Unstable modes

One can expect there to be normal-mode solutions in the form of

$$\psi'_n = \mathrm{Real}\left(\Psi_n e^{i(kx + \ell y - \omega t)} \right), \tag{5.56}$$

with analogous expressions for the other dependent variables, because the linear partial differential equations in (5.55) have constant coefficients. Inserting (5.56) into (5.55) and factoring out the exponential function gives

$$(C-U)\left[K^2\Psi_1 + \frac{1}{2R^2}(\Psi_1 - \Psi_2)\right] + \left[\beta + \frac{U}{R^2}\right]\Psi_1 = 0$$

$$(C+U)\left[K^2\Psi_2 - \frac{1}{2R^2}(\Psi_1 - \Psi_2)\right] + \left[\beta - \frac{U}{R^2}\right]\Psi_2 = 0 \qquad (5.57)$$

for $C = \omega/k$ and $K^2 = k^2 + \ell^2$. Redefine the variables by transforming the layer amplitudes into vertical modal amplitudes by (5.36):

$$\tilde{\Psi}_0 \equiv \frac{1}{2}(\Psi_1 + \Psi_2)$$

$$\tilde{\Psi}_1 \equiv \frac{1}{2}(\Psi_1 - \Psi_2). \qquad (5.58)$$

These are the barotropic and baroclinic vertical modes, respectively. The linear combinations of layer coefficients are the vertical eigenfunctions associated with $R_0 = \infty$ and $R_1 = R$ from (5.39). Now take the sum and difference of the equations in (5.57) and substitute (5.58) to obtain the following modal amplitude equations:

$$[CK^2 + \beta]\tilde{\Psi}_0 - UK^2\tilde{\Psi}_1 = 0$$

$$[C(K^2 + R^{-2}) + \beta]\tilde{\Psi}_1 - U(K^2 - R^{-2})\tilde{\Psi}_0 = 0. \qquad (5.59)$$

For the special case with no mean flow, $U = 0$, the first equation in (5.59) is satisfied for $\tilde{\Psi}_0 \neq 0$ only if

$$C = C_0 = -\frac{\beta}{K^2}. \qquad (5.60)$$

$\tilde{\Psi}_0$ is the barotropic vertical modal amplitude, and this relation is identical to the dispersion relation for barotropic Rossby waves with an infinite deformation radius (Section 3.1.2). The second equation in (5.59) with $\tilde{\Psi}_1 \neq 0$ implies that if

$$C = C_1 = -\frac{\beta}{K^2 + R^{-2}}. \qquad (5.61)$$

$\tilde{\Psi}_1$ is the baroclinic vertical modal amplitude, and the expression for C is the same as the dispersion relation for baroclinic Rossby waves with finite deformation radius, R (Section 4.7).

When $U \neq 0$, (5.59) has non-trivial modal amplitudes, $\tilde{\Psi}_0$ and $\tilde{\Psi}_1$, only if the determinant for their second-order system of linear algebraic equations vanishes, namely,

$$[CK^2 + \beta][C(K^2 + R^{-2}) + \beta] - U^2K^2[K^2 - R^{-2}] = 0. \qquad (5.62)$$

This is the general dispersion relation for this normal-mode problem.

To understand the implications of (5.62) with $U \neq 0$, first consider the case of $\beta = 0$. Then the dispersion relation can be rewritten as

$$C^2 = U^2 \frac{K^2 - R^{-2}}{K^2 + R^{-2}}. \tag{5.63}$$

For all $KR < 1$ (i.e., the long waves), $C^2 < 0$. This implies that C is purely imaginary with an exponentially growing modal solution (i.e., an instability) and a decaying one, proportional to

$$e^{-ikCt} = e^{k \, \mathrm{Imag}[C]t}.$$

This behavior is a baroclinic instability for a mean flow with shear only in the vertical direction.

For $U, \beta \neq 0$, the analogous condition for C having a non-zero imaginary part is when the discriminant of the quadratic dispersion relation (5.62) is negative, i.e., $\mathcal{P} < 0$ for

$$\mathcal{P} \equiv \beta^2 (2K^2 + R^{-2})^2 - 4(\beta^2 K^2 - U^2 K^4 (K^2 - R^{-2})) (K^2 + R^{-2})$$
$$= \beta^2 R^{-4} + 4U^2 K^4 (K^4 - R^{-4}). \tag{5.64}$$

Note that β tends to stabilize the flow because it acts to make $\mathcal{P}$ more positive and thus reduces the magnitude of $\mathrm{Imag}\,[C]$ when $\mathcal{P}$ is negative. Also note that in both (5.63) and (5.64) the instability is equally strong for either sign of U (i.e., eastward or westward vertical shear).

The smallest value for $\mathcal{P}(K)$ occurs when

$$0 = \frac{\partial \mathcal{P}}{\partial K^4} = 4U^2(K^4 - R^{-4}) + 4U^2 K^4, \tag{5.65}$$

or

$$K = \frac{1}{2^{1/4} R}. \tag{5.66}$$

At this K value, the value for $\mathcal{P}$ is

$$\mathcal{P} = \beta^2 R^{-4} - U^2 R^{-8}. \tag{5.67}$$

Therefore, a necessary condition for instability is

$$U > \beta R^2. \tag{5.68}$$

From (5.52) this condition is equivalent to the mean potential vorticity gradients, $d_y \overline{q}_{\mathrm{QG},n}$, having opposite signs in the two layers,

$$\frac{d\overline{q}_{\mathrm{QG},1}}{dy} \cdot \frac{d\overline{q}_{\mathrm{QG},2}}{dy} < 0.$$

The instability requirement for a sign change in the mean (potential) vorticity gradient is similar to the Rayleigh criterion for barotropic vortex instability (Section 3.3.1), and, not surprisingly, a Rayleigh criterion may also be derived for quasigeostrophic baroclinic instability.

Further analysis of $\mathcal{P}(K)$ shows other conditions for instability:

- $KR < 1$ is necessary (and it is also sufficient when $\beta = 0$);
- $U > \frac{1}{2}\beta(R^{-4} - K^4)^{-1/2} \to \infty$ as $K \to R^{-1}$ from below; and
- $U > \frac{1}{2}\beta K^{-2} \to \infty$ as $K \to 0$ from above.

These relations support the *regime diagram* in Fig. 5.5 for baroclinic instability. For any $U > \beta R^2$, there is a perturbation length scale for the most unstable mode that is somewhat greater than the baroclinic deformation radius. Short waves ($K^{-1} < R$) are stable, and very long waves ($K^{-1} \to \infty$) are stable through the influence of β.

When $\mathcal{P} < 0$, the solution to (5.62) is

$$C = -\frac{\beta(2K^2 + R^{-2})}{2K^2(K^2 + R^{-2})} \pm \frac{i\sqrt{-\mathcal{P}}}{2K^2(K^2 + R^{-2})}. \qquad (5.69)$$

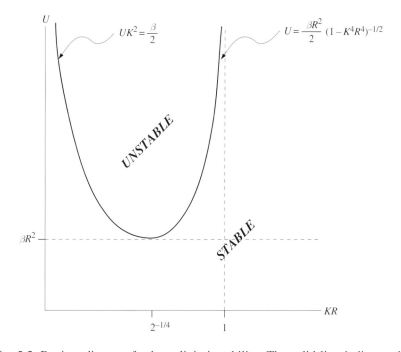

Fig. 5.5. Regime diagram for baroclinic instability. The solid line indicates the marginal stability curve as a function of the mean vertical shear amplitude, U, and perturbation wavenumber, K, for $\beta \neq 0$. The vertical dashed line is the marginal stability curve for $\beta = 0$.

Thus the zonal phase propagation for unstable modes (i.e., the real part of C) is to the west. From (5.69),

$$-\frac{\beta}{K^2} < \text{Real}\,[C] < -\frac{\beta}{K^2 + R^{-2}}. \tag{5.70}$$

The unstable-mode phase speed lies in between the barotropic and baroclinic Rossby-wave speeds in (5.60) and (5.61). This result is demonstrated by substituting the first term in (5.69) for Real $[C]$ and factoring $-\beta/K^2$ from all three expressions in (5.70). These steps yield

$$1 \geq \frac{1 + \mu/2}{1 + \mu} \geq \frac{1}{1 + \mu} \tag{5.71}$$

for $\mu = (KR)^{-2}$. These inequalities are obviously true for all $\mu \geq 0$.

5.2.2 Upshear phase tilt

From (5.59),

$$\tilde{\Psi}_1 = \frac{C + \beta K^{-2}}{U} \tilde{\Psi}_0$$

$$= \left| \frac{C + \beta K^{-2}}{U} \right| e^{i\theta} \tilde{\Psi}_0, \tag{5.72}$$

where θ is the phase angle for $(C + \beta K^{-2})/U$ in the complex plane. Since

$$\text{Real}\left[\frac{C + \beta K^{-2}}{U} \right] > 0 \tag{5.73}$$

from (5.70), and

$$\text{Imag}\left[\frac{C + \beta K^{-2}}{U} \right] = \frac{\text{Imag}\,[C]}{U} > 0 \tag{5.74}$$

for growing modes (with Real $[-ikC] > 0$, i.e., Imag $[C] > 0$), then $0 < \theta < \pi/2$ in westerly wind shear ($U > 0$). As shown in Fig. 5.6 this implies that $\tilde{\psi}_1$ has its pattern shifted to the west relative to $\tilde{\psi}_0$, by an amount less than a quarter wavelength. A graphical addition and subtraction of $\tilde{\psi}_1$ and $\tilde{\psi}_0$ according to (5.58) is shown in Fig. 5.6. It indicates that the layer ψ_1 has its pattern shifted to the west relative to ψ_2, by an amount less than a half wavelength. Therefore, upper-layer disturbances are shifted to the west relative to lower-layer ones; i.e., they are tilted upstream with respect to the mean shear direction (Fig. 5.7). This feature is usually evident on weather maps during the amplifying phase for mid-latitude cyclonic synoptic storms and is often used as a synoptic analyst's rule of thumb.

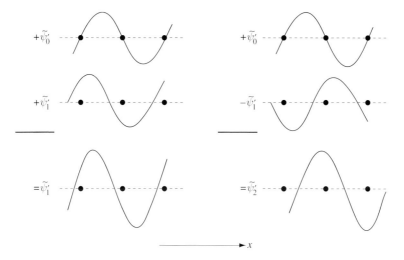

Fig. 5.6. Modal and layer phase relations for the perturbation streamfunction, $\psi'(x, t)$, in baroclinic instability for a two-layer fluid. This plot exhibits graphical addition: in each column the modal curves in the top two rows are added together to obtain the respective layer curves in the bottom row.

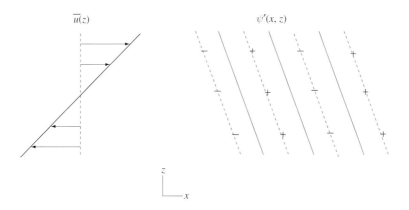

Fig. 5.7. Up shear phase tilting for the perturbation streamfunction, $\psi'(x, z)$, in baroclinic instability for a continuously stratified fluid.

5.2.3 Eddy heat flux

Now calculate the poleward *eddy heat flux*, $\overline{v'T'}$ (disregarding the conversion factor, $\rho_0 c_p$, between temperature and heat; Section 2.1.2). The heat flux is analogous to a Reynolds stress (Section 3.4) as a contributor to the dynamical balance relations for the equilibrium state, except it appears in the mean heat equation rather than the mean momentum equation. Here $v' = \partial_x \psi'$, and the temperature fluctuation is associated with the interfacial displacement as in (5.9),

$$T' = \frac{b'}{\alpha g} = \frac{1}{\alpha g}\frac{\partial \phi'}{\partial z} = \frac{f}{\alpha g}\frac{\partial \psi'}{\partial z} = \frac{2f}{\alpha g H}(\psi'_1 - \psi'_2) = \frac{4f}{\alpha g H}\tilde{\psi}'_1,$$

with all the proportionality constants positive in the northern hemisphere. Suppose that at some time the modal fields have the (x, z) structure,

$$\tilde{\psi}'_1 = A_1 \sin[kx + \theta]$$

$$\tilde{\psi}'_0 = A_0 \sin[kx] \tag{5.75}$$

for $A_0, A_1 > 0$ and $0 < \theta < \pi/2$ (Fig. 5.6). Then

$$\tilde{v}'_1 = A_1 k \cos[kx + \theta]$$

$$\tilde{v}'_0 = A_0 k \cos[kx]. \tag{5.76}$$

The layer velocities, v'_n, are proportional to the sum of $\tilde{v}'_0$ and $\pm\tilde{v}'_1$ in the upper and lower layers, respectively, as in (5.37). Therefore the modal heat fluxes are

$$\overline{\tilde{v}'_1 T'} \equiv \frac{k}{2\pi}\int_0^{2\pi} dx\, \tilde{v}'_1 T'$$

$$\propto \int_0^{2\pi} dx \sin[kx + \theta]\cos[kx + \theta] = 0$$

$$\overline{\tilde{v}'_0 T'} \equiv \frac{k}{2\pi}\int_0^{2\pi} dx\, \tilde{v}'_0 T'$$

$$\propto \frac{k}{2\pi}\int_0^{2\pi} dx \sin[kx + \theta]\cos[kx] = \frac{\sin[\theta]}{2} \tag{5.77}$$

with positive proportionality constants. Since each v'_n has a positive contribution from $\tilde{v}'_0$, the interfacial heat flux, $\overline{v'T'}$, is proportional to $\overline{\tilde{v}'_0 T'}$, and it is therefore positive, $\overline{v'T'} > 0$. The sign of $\overline{v'T'}$ is directly related to the range of values for θ, i.e., to the upshear vertical phase tilt (Section 5.2.2).

5.2.4 Effects on the mean flow

The non-zero eddy heat flux for baroclinic instability implies there is an eddy–mean interaction. A mean energy balance is derived similarly to the energy conservation relation (5.16) by manipulation of the mean momentum and thickness equations. The result has the following form in the present context:

$$\frac{d}{dt}\overline{E} = \cdots + \iint dx\,dy\, g'_1\overline{v'\eta'}\frac{d\overline{\eta}}{dy}, \tag{5.78}$$

where the dots refer to any non-conservative processes (here unspecified) and the mean-flow energy is defined by

$$\overline{E} = \iint dx\, dy\, \frac{1}{2}\left(\overline{h}_1 \overline{\mathbf{u}}_1^2 + \overline{h}_2 \overline{\mathbf{u}}_2^2 + g_I' \overline{\eta}^2\right). \tag{5.79}$$

Analogous to (3.100) for barotropic instability, there is a baroclinic energy conversion term here that generates fluctuation energy by removing it from the mean energy when the eddy flux, $\overline{v'\eta'}$, has the opposite sign to the mean gradient, $d_y \overline{\eta}$. Since η is proportional to T in a layered model, this kind of conversion occurs when $\overline{v'T'} > 0$ and $d_y \overline{T} < 0$ (as shown in Section 5.2.3).

The eddy–mean interaction cannot be fully analyzed in the spatially homogeneous formulation of this section, implicit in the horizontally periodic eigenmodes (5.56). It is the divergence of the eddy heat flux that causes changes in the mean temperature gradient,

$$\frac{\partial \overline{T}}{\partial t} = \cdots - \frac{\partial}{\partial y}\overline{v'T'},$$

and the divergence is zero in a homogeneous flow. Thus, a more complete interpretation of the role of eddies in the general circulation requires an extension to inhomogeneous flows, such as the tropospheric westerly jet that has its maximum speed at a middle latitude, $\approx 45°$.

The poleward heat flux in baroclinic instability tends to weaken the mean state by transporting warm air fluctuations into the region on the poleward side of the jet with its associated mean-state cold air (NB, Fig. 5.1). Equation (5.78) shows that the mean circulation loses energy as the unstable fluctuations grow in amplitude: the mean meridional temperature gradient (hence the mean geostrophic shear) is diminished by the eddy heat flux, and part of the mean available potential energy associated with the meridional temperature gradient is converted into eddy energy. The mid-latitude atmospheric climate is established as a balance between the acceleration of the westerly Jet Stream by Equator-to-pole differential radiative heating and the limitation of the vertical shear strength of the jet by the unstable eddies that transport heat between the equatorial heating and polar cooling zones.

A similar interpretation can be made for the zonally directed *Antarctic Circumpolar Current* (ACC) in the ocean (Fig. 6.11). In the wind-driven ACC, the more natural dynamical characterization is in terms of the mean momentum balance rather than the mean heat balance, although these two balances must be closely related because of thermal wind balance. A mean eastward wind stress beneath the westerly winds drives a surface-intensified, eastward mean current that is baroclinically unstable and generates eddies that transfer momentum vertically. This eddy momentum transfer has to be balanced against a bottom turbulent drag and/or topographic form stress (a pressure force against the solid bottom topography; Section 5.3.3). The eddies also transport heat southward (poleward in the southern hemisphere), balanced by the advective heat flux caused by the mean, ageostrophic, secondary circulation in the meridional (y, z) plane, such that there is no net heat flux by their combined effects.

In these descriptions for the baroclinically unstable westerly winds and ACC, note two important ideas about the dynamical maintenance of a mean zonal flow.

- An equivalence between horizontal heat flux and vertical momentum flux for quasi-geostrophic flows. The latter process is referred to as *isopycnal form stress*. It is analogous to topographic form stress except that the relevant material surface is an isopycnal in the fluid interior instead of the solid bottom. Isopycnal form stress is not the vertical Reynolds stress, $\langle u'w' \rangle$, which is much weaker than the isopycnal form stress for quasigeostrophic flows because w' is so weak (Section 4.7).
- The existence of a mean secondary circulation in the (y, z) plane, perpendicular to the main zonal flow, associated with the eddy heat and momentum fluxes whose mean meridional advection of heat may partly balance the poleward eddy heat flux. This is called the *Deacon Cell* for the ACC and the *Ferrel Cell* for the westerly winds. It also is referred to as the *meridional overturning circulation*.

In the next section these behaviors are illustrated in an idealized problem for the statistical equilibrium state of an inhomogeneous zonal jet, and the structures of the mean flow, eddy fluxes, and secondary circulation are examined.

5.3 Turbulent baroclinic zonal jet

5.3.1 Posing the jet problem

Consider a computational solution for an N-layer quasigeostrophic model (Section 5.1.2) that demonstrates the phenomena discussed at the end of the previous section. The problem could be formulated for a zonal jet forced either by a meridional heating gradient (e.g., the mid-latitude westerly winds in the atmosphere) or by a zonal surface stress (e.g., the ACC in the ocean). The latter is adopted because it embodies the essentially adiabatic dynamics in baroclinic instability and its associated eddy–mean interactions. It is an idealized model for the ACC, neglecting both the actual wind and basin geographies and the diabatic surface fluxes and interior mixing. For historical reasons (McWilliams & Chow, 1981), a solution is presented here with N set to 3; this is an N value larger by one than the minimum vertical resolution, $N = 2$, needed to represent baroclinic instability (Section 5.2).

This idealization for the ACC is as an adiabatic, quasigeostrophic, zonally periodic jet driven by a broad, steady, zonal surface wind stress. The flow environment is a southern-hemisphere, β-plane approximation to the Coriolis frequency and has an irregular bottom topography (which can be included in the bottom layer of an N-layer model analogous to its inclusion in a shallow-water model; Section 4.1). The mean stratification is specified so that the baroclinic deformation radii, the R_m from (5.46), are much smaller than both the meridional wind scale, L_τ, and the ACC meridional velocity scale, L; the latter are also specified to be comparable to the domain width, L_y (i.e., $R_m \ll L, L_\tau, L_y \forall\ m \geq 1$). This problem configuration is sketched in Fig. 5.8. Another important scale is $L_\beta = \sqrt{V/\beta}$, with V a typical velocity associated with either the mean or eddy currents. This is the Rhines scale (Section 4.8.1). In both the ACC and this idealized solution, L_β is somewhat smaller than L_y, although not by much. The domain is a meridionally bounded, zonally periodic channel with solid side boundary conditions of no normal flow and zero lateral stress. However, since the wind stress decays in amplitude away from the channel center toward the walls, as do both the mean zonal jet and its eddies, the meridional boundaries do not play a significant role in the solution behavior (cf., the essential role of a western boundary current in a wind gyre; Section 6.2). The resting layer depths are chosen to have the values, $H_n = [500, 1250,\ 3250]$ m. They are unequal in size, as is commonly done to represent the fact that mean stratification, $\mathcal{N}(z)$, increases in the upper ocean. The reduced gravity values, $g'_{n+0.5}$, are then chosen so that the associated deformation radii

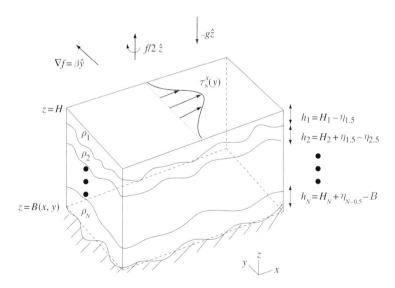

Fig. 5.8. Posing the zonal jet problem for an N-layer model for a rotating, stratified fluid on the β plane with surface wind stress and bottom topography. The black dots indicate deleted intermediate layers for $n = 3$ to $N - 1$.

are $R_m = [\infty, 32, 15]$ km after solving the eigenvalue problem in Section 5.1.3, and these values are similar to those for the real ACC. Both of the baroclinic R_m (i.e., $m \geq 1$) values are small compared to the chosen channel width of $L_y = 1000$ km.

The wind stress accelerates a zonal flow. To have any chance of arriving at an equilibrium state, the problem must be posed to include non-conservative terms, e.g., with horizontal and vertical eddy viscosities (Section 3.5), ν_h and ν_v [m² s⁻¹], and/or a *bottom-drag* damping coefficient, ϵ_{bot} [m s⁻¹]. Non-conservative terms have not been discussed very much so far, and they will merely be stated here in advance of the more extensive discussion in Chapter 6. In combination with the imposed zonal surface wind stress, $\tau_s^x(y)$, these non-conservative quantities are expressed in the non-conservative horizontal force as

$$\mathbf{F}_1 = \frac{\tau_s^x}{\rho_0 H_1}\hat{\mathbf{x}} + \nu_h \nabla^2 \mathbf{u}_1 + \frac{2\nu_v}{H_1}\left(\frac{\mathbf{u}_2 - \mathbf{u}_1}{H_1 + H_2}\right)$$

$$\mathbf{F}_n = \nu_h \nabla^2 \mathbf{u}_n + \frac{2\nu_v}{H_n}\left(\frac{\mathbf{u}_{n+1} - \mathbf{u}_n}{H_n + H_{n+1}} + \frac{\mathbf{u}_{n-1} - \mathbf{u}_n}{H_n + H_{n-1}}\right), \quad 2 \leq n \leq N-1,$$

$$\mathbf{F}_N = \nu_h \nabla^2 \mathbf{u}_N + \frac{2\nu_v}{H_N}\left(\frac{\mathbf{u}_{N-1} - \mathbf{u}_N}{H_N + H_{N-1}}\right) - \frac{\epsilon_{bot}}{H_N}\mathbf{u}_N. \quad (5.80)$$

The wind stress is a forcing term in the upper layer ($n = 1$); the bottom drag is a damping term in the bottom layer ($n = N$); the horizontal eddy viscosity multiplies a second-order horizontal Laplacian operator on $\mathbf{u}_n$, analogous to molecular viscosity (Section 2.1.2); and the vertical eddy viscosity multiplies a finite-difference approximation to the analogous second-order vertical derivative operating on $\mathbf{u}(z)$. In the quasigeostrophic potential-vorticity equations (5.26) and (5.27), these non-conservative terms enter as the force curl, $\mathcal{F}_n$. The potential-vorticity equations are solved for the geostrophic layer streamfunctions, ψ_n, and the velocities in (5.80) are evaluated geostrophically.

The top and bottom boundary stress terms appear as equivalent *body forces* in the layers adjacent to the boundaries. The underlying concept for the boundary stress terms is that they are conveyed to the fluid interior through turbulent boundary layers, called *Ekman layers*, whose thickness is much smaller than the layer thickness of the model. So the vertical flow structure within the Ekman layers cannot be explicitly resolved in the layered model. Instead the Ekman layers are conceived of as thin sub-layers embedded within the $n = 1$ and N resolved layers, and they cause near-boundary vertical velocities, called *Ekman pumping*, at the interior layer interfaces closest to the boundaries.

In turn the Ekman pumping causes vortex stretching in the rest of the resolved layer and thereby acts to modify both the thickness and potential vorticity of the layer. The boundary stress terms in (5.80) have the net effects summarized here; the detailed functioning of a turbulent boundary layer is explained in Chapter 6.

If the eddy diffusion parameters are large enough (i.e., the effective Reynolds number, Re, is small enough), they can viscously support a steady, stable, laminar jet in equilibrium against the acceleration by the wind stress. However, for smaller diffusivity values – as certainly required for geophysical plausibility – the accelerating jet will become unstable before it reaches a viscous stationary state. A *bifurcation sequence* of successive instabilities with increasing Re values can be mapped out, but most geophysical jets are well past this *transition regime* in Re. The jets can reach an equilibrium state only through coexistence with a state of *fully developed turbulence* comprised by the geostrophic, mesoscale eddies generated by the mean jet instabilities. Accordingly, the values for ν_h and ν_v in the computational solution are chosen to be small in order to yield fully developed turbulence. The most important type of jet instability for broad baroclinic jets, with $L_y \gg R_1$, is baroclinic instability (Section 5.2). In fully developed turbulence the eddies grow by instability of the mean currents, and they cascade the variance of the fluctuations from their generation scale to the dissipation scale (cf., Section 3.7). In equilibrium the average rates for these processes must be equal. In turn, the turbulent eddies limit and reshape the mean circulation (as described at the end of Section 5.2.4) in an eddy–mean interaction.

5.3.2 Equilibrium velocity and buoyancy structure

First consider the flow patterns and the geostrophically balanced buoyancy field for fully developed turbulence in the statistical equilibrium state that develops during a long-time integration of the three-layer quasigeostrophic model. The instantaneous $\psi_n(x, y)$ and $q_{QG,n}(x, y)$ fields are shown in Fig. 5.9, and the $T(x, y) = b(x, y)/\alpha g$ and w fields are shown in Fig. 5.10 (NB, $f < 0$ since this is for the southern hemisphere). Note the strong, narrow, meandering jet in the upper ocean and the weaker, broader flow in the abyssal ocean. The instantaneous centerline for the jet is associated with a continuous front in b, a broken front in q, and extrema in w alternating in sign within the eddies and along the meandering jet axis.

The mean state is identified by an overbar defined as an average over (x, t). The domain for each of these coordinates is taken to be infinite, consistent with our interpretive assumptions of zonal homogeneity and stationarity for the problem posed in Section 5.3.1, even though the domain is necessarily finite

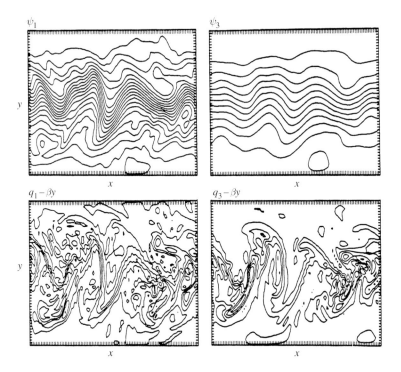

Fig. 5.9. Instantaneous horizontal patterns for streamfunction, ψ_n, and quasi-geostrophic potential vorticity, $q_{QG,n}$ (excluding its βy term), in the upper- and lower-most layers in a zonal-jet solution with $N = 3$. The ψ contour interval is 1.5×10^{-4} m^2 s^{-1} and the q contour intervals are 2.5 ($n = 1$) and 0.25 ($n = 3$) $\times 10^{-4}$ s^{-1} (adapted from McWilliams & Chow, 1981).

(but large, for statistical accuracy) in a computational solution. The combination of periodicity and translational symmetry (literal or statistical) for the basin shape, wind stress, and topography is a common finite-extent approximation to homogeneity. The mean geostrophic flow is a surface intensified zonal jet, $\overline{u}_n(y)$, sketched in Fig. 5.11. This jet is in hydrostatic, geostrophic balance with the dynamic pressure, $\overline{\phi}_m$; streamfunction, $\overline{\psi}_n$; layer thickness, $\overline{h}_n$; interfacial elevation anomaly, $\overline{\eta}_{n+0.5}$; and interfacial buoyancy anomaly, $\overline{b}_{n+0.5}$ – each defined in Section 5.2 and sketched in Fig. 5.12.

The 3D mean circulation is $(\overline{u}, \overline{v}_a, \overline{w})$. Only the zonal component is in geostrophic balance. The meridional flow cannot be in geostrophic balance because there can be no mean zonal pressure gradient in a zonally periodic channel, and vertical velocity is never geostrophic by definition (Section 2.4.2). Thus both components of the mean velocity in the meridional plane (i.e., the meridional overturning circulation that is an idealized form of the Deacon Cell; Section 5.2.4) is ageostrophic and thus weaker than $\overline{u}$ by $\mathcal{O}(Ro)$.

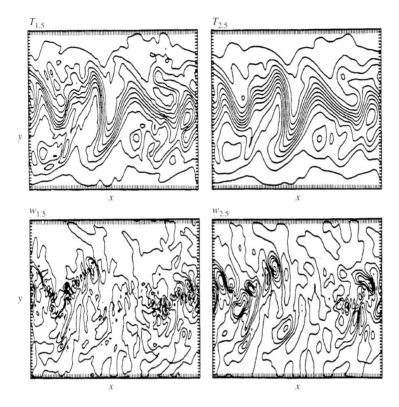

Fig. 5.10. Instantaneous horizontal patterns for temperature, $T_{n+0.5}$, and vertical velocity, $w_{n+0.5}$, at the upper and lower interior interfaces in a quasigeostrophic zonal-jet solution with $N = 3$. The T and w contour intervals are 0.4 and 0.1 K and 10^{-4}m s^{-1} and 0.5×10^{-4}m s^{-1} at the upper and lower interfaces, respectively (adapted from McWilliams & Chow, 1981).

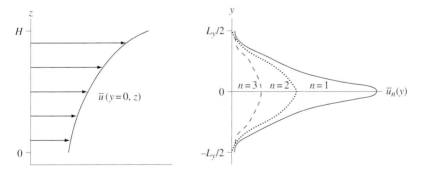

Fig. 5.11. Sketch of the time-mean zonal flow, $\bar{u}(y, z)$, in the equilibrium jet problem with $N = 3$: (left) vertical profile in the middle of the channel and (right) meridional profiles in different layers. Note the intensification of the mean jet toward the surface and the middle of the channel.

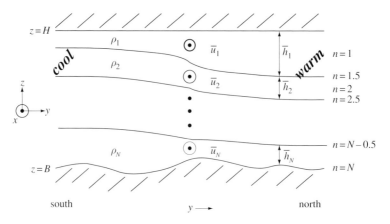

Fig. 5.12. Sketch of a meridional cross-section for the time-mean zonal jet, the layer thickness, the density, and the buoyancy anomaly.

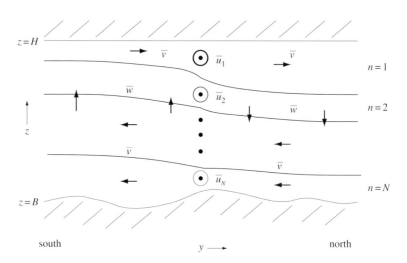

Fig. 5.13. Sketch of the time-mean, meridional overturning circulation (i.e., Deacon Cell) for the zonal jet, overlaid on the mean zonal jet and layer thickness.

The overturning circulation is sketched in Fig. 5.13. Because of the zonal periodicity, its component velocities satisfy a 2D continuity relation pointwise (cf., (4.112)):

$$\frac{\partial \overline{v}_a}{\partial y} + \frac{\partial \overline{w}}{\partial z} = 0. \tag{5.81}$$

(A 2D zonally averaged continuity equation also occurs for the meridional over-turning circulation with solid boundaries in x, e.g., as in Section 6.2.) This relation will be further examined in the context of the layer mass balance (Section 5.3.5).

The meridional profiles of mean and eddy-variance quantities in Figs. 5.14 and 5.15 show the following features:

- an eastward jet that increases its strength with height (cf., the atmospheric westerly winds);
- geostrophically balancing temperature gradients (with cold water on the poleward side of the jet);
- opposing potential vorticity gradients in the top and bottom layers (i.e., satisfying a Rayleigh necessary condition for baroclinic instability; Section 5.2.1);
- a nearly uniform $q_{\text{QG},2}(y)$ in the middle layer (NB this is a consequence of the eddy mixing of q_{QG}, sometimes called *potential-vorticity homogenization*, in a layer without significant non-conservative forces; Section 5.3.4);
- mean upwelling on the poleward side of the jet and downwelling on the equatorward side (i.e., a Deacon Cell, the overturning secondary circulation in the meridional plane, with an equatorward surface branch that is Ekman-layer transport due to eastward wind stress in the southern hemisphere; Chapter 6);
- eddy variance profiles for ψ', u', v', and $T' = b'/\alpha g$ that decay both meridionally and vertically away from the jet core.

To understand how this equilibrium is dynamically maintained, the mean dynamical balances for various quantities will be analyzed in Sections 5.3.3–5.3.6. In each case the eddy flux makes an essential contribution (cf., Section 3.4).

5.3.3 Zonal momentum balance

What is the mean zonal momentum balance for the statistical equilibrium state? Its most important part (i.e., neglecting mean eddy diffusion) is

$$\frac{\partial \overline{u}_n}{\partial t} = 0 \text{ for all } n$$

$$\approx \frac{\tau_s^x}{\rho_0 H_1} - \frac{\mathcal{D}_{1.5}}{H_1} - \frac{\partial \mathcal{R}_1}{\partial y}, \quad n = 1$$

$$\approx \left(\frac{\mathcal{D}_{n-0.5} - \mathcal{D}_{n+0.5}}{H_n} \right) - \frac{\partial \mathcal{R}_n}{\partial y}, \quad 2 \leq n \leq N-1$$

$$\approx -\frac{\epsilon_{\text{bot}}}{H_N} \overline{u}_N + \left(\frac{\mathcal{D}_{N-0.5} - \mathcal{D}_{\text{bot}}}{H_N} \right) - \frac{\partial \mathcal{R}_N}{\partial y}, \quad n = N \qquad (5.82)$$

with eddy fluxes, $\mathcal{R}$ and $\mathcal{D}$, defined in (5.84) and (5.87), respectively. All contributions to the mean momentum balance from mean advective, geostrophic-Coriolis, and pressure-gradient forces are absent at the leading order of expansion in *Ro* due

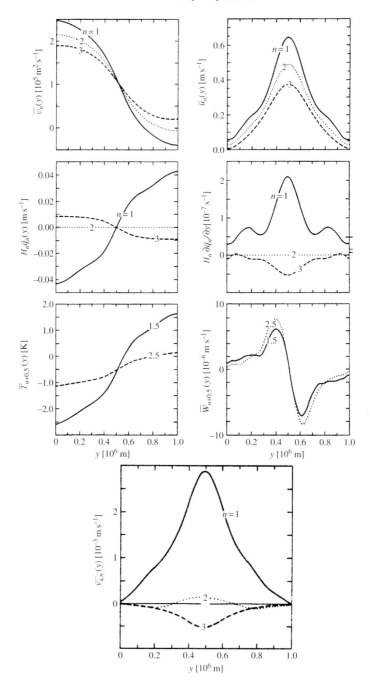

Fig. 5.14. Time-mean meridional profiles for a quasigeostrophic zonal-jet solution with $N = 3$. The panels (clockwise from the upper left) are for streamfunction, geostrophic zonal velocity, potential vorticity gradient, vertical velocity, ageostrophic meridional velocity, temperature, and potential vorticity (adapted from McWilliams & Chow, 1981).

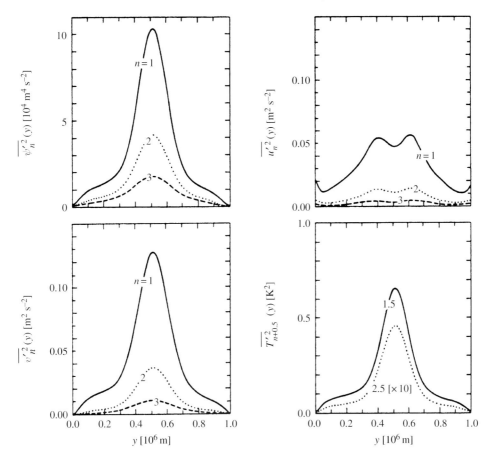

Fig. 5.15. Meridional eddy variance profiles for a quasigeostrophic zonal-jet solution with $N = 3$. The panels (clockwise from the upper left) are for stream-function, geostrophic zonal velocity, temperature, and geostrophic meridional velocity (adapted from McWilliams & Chow, 1981).

to the (x, t)-averaging. Also, a substitution has been made for the ageostrophic Coriolis force,

$$f\overline{v}_{a,n} = -\frac{f_0}{H_n}\overline{(h_n - H_n)v_n}, \tag{5.83}$$

in favor of the eddy mass fluxes, $\overline{(h_n - H_n)v_n}$, that are comprised of the interfacial quantities, $\mathcal{D}_{n+0.5}$. (This substitution relation comes from the mean layer thickness equation (5.95) in Section 5.3.5.) The reason for making the substitution for the mean quantity, $\overline{v}_{a,n}$, in favor of the eddy flux, $\mathcal{D}$, is to emphasize the central role for the turbulent eddies in maintaining a force balance in the statistical equilibrium dynamics. In (5.82) the mean wind and bottom stress are readily identifiable. The remaining terms are eddy flux divergences of two different types.

The more important type – because baroclinic instability is the more important eddy generation process for a broad, baroclinic jet – is the vertical divergence of $\mathcal{D}$, an eddy form stress defined in (5.87). A different type of eddy flux in (5.82), secondary in importance to $\mathcal{D}$ for this type of jet flow, is the horizontal divergence of the horizontal Reynolds stress,

$$\mathcal{R}_n = \overline{u_{g,n} v_{g,n}} \tag{5.84}$$

(cf., Section 3.4, where $\mathcal{R}$ is the important eddy flux for a barotropic flow).

First focus on the momentum balance within a single vertical column. If the contributions from $\mathcal{R}$ are temporarily neglected, a vertical integral of (5.82), symbolically expressed here in a continuous vertical coordinate, is

$$\frac{\partial}{\partial t} \int_{\text{bot}}^{\text{top}} \overline{u} \, dz \, [= 0] \approx \frac{\tau_s^x}{\rho_0} - [\mathcal{D}_{\text{bot}} + \epsilon_{\text{bot}} \overline{u}_{\text{bot}}]. \tag{5.85}$$

For $\tau_s^x > 0$ (eastward wind stress on the ocean due to westerly surface winds), opposing bottom contributions are necessary from $\overline{u}_{\text{bot}} > 0$ and/or $\mathcal{D}_{\text{bot}} > 0$. The latter, the zonal *topographic form stress*, generally dominates over the bottom drag due to the turbulent bottom boundary layer (Chapter 6) in both the westerly winds and the ACC. Without topographic form stress, the turbulent drag over a flat surface is so inefficient with realistic values for ϵ_{bot} (representing the effect of bottom boundary layer turbulence) that an unnaturally large bottom velocity and zonal transport are required to reach equilibrium in the zonal momentum balance.

The definition for $\mathcal{D}_{\text{bot}}$ is

$$\mathcal{D}_{\text{bot}} = \overline{\phi_N \frac{\partial B}{\partial x}} = -f_0 \, \overline{v_{g,N} B}. \tag{5.86}$$

$B(x, y)$ is the anomalous bottom height relative to its mean depth. The second relation in (5.86) involves a zonal integration by parts and the use of geostrophic balance. $\mathcal{D}_{\text{bot}}$ has an obvious interpretation (Fig. 5.16) as the integrated horizontal pressure force pushing against the bottom slopes. $z = B$ is a material surface, so it separates the fluid above from the land below. If the pressure differs on different sides of a bump, then a force acts to push the bump sideways, and in turn the bump pushes back on the fluid. This force is analogous to the lift (vertical) and drag (horizontal) forces exerted by an inviscid flow past an airplane wing. Notice that $\overline{v}_a$ does not contribute to $\mathcal{D}_{\text{bot}}$ both because it is weak (i.e., ageostrophic) and because it is the same on either side of a bottom slope. Similarly the v' from *transient eddies* does not contribute to bottom form stress because B is time-independent and the time average of their product is zero. The only part of

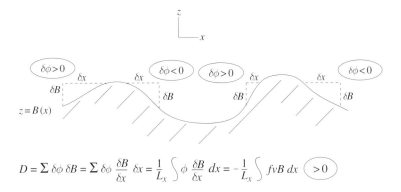

$$D = \Sigma \, \delta\phi \, \delta B = \Sigma \, \delta\phi \, \frac{\delta B}{\delta x} \, \delta x = \frac{1}{L_x} \int \phi \, \frac{\delta B}{\delta x} \, dx = -\frac{1}{L_x} \int fvB \, dx \quad \boxed{>0}$$

Fig. 5.16. Topographic form stress (cf., (5.86)) in the zonal direction. The difference of pressure on either side of an extremum in the bottom elevation, B, contributes to a zonally averaged force, D, that can be expressed either as the product of the pressure anomaly and the zonal bottom slope or, for geostrophic flow, as minus the product of the meridional velocity and the bottom elevation.

the flow contributing to the mean bottom form stress is the *standing eddies*, the time-mean deviation from the (x, t)-mean flow.

In order to sustain the situation with equal and opposite surface and bottom zonal stresses, there must also be a mechanism that transmits zonal stress downward through the interior layers and across the layer interfaces. This process is represented by $\mathcal{D}_{n+0.5}$, defined by

$$\mathcal{D}_{n+0.5} = \overline{\phi_{n+0.5} \frac{\partial \eta_{n+0.5}}{\partial x}} = -f_0 \, \overline{v_{g,n+0.5} \eta_{n+0.5}}. \tag{5.87}$$

There is an obvious isomorphism with $\mathcal{D}_{\mathrm{bot}}$, except the relevant material surface is now the moving interface rather than the stationary bottom. For the layer n above the interface $n+0.5$, $\mathcal{D}_{n+0.5}$ is the force exerted by the layer on the interface, and for the layer $n+1$ below it is the reverse. Furthermore, if $v_{n+0.5}$ is defined as a simple average, $(v_n + v_{n+1})/2$, and (5.20) is used to replace η by ψ, then

$$\mathcal{D}_{n+0.5} = -\frac{f_0^2}{g'_{n+0.5}} \, \overline{v_{g,n} \psi_{n+1}}. \tag{5.88}$$

Note that both transient and standing eddies can contribute to the isopycnal form stress.

If $\mathcal{R}$ is truly negligible in (5.82), then all the interior $\mathcal{D}_{n+0.5}$ are equal to τ_s^x. More generally, in the jet center in each interior layer, the mean zonal flow is accelerated by the difference between the isopycnal form stress at the interfaces (isopycnal surfaces) above and below the layer, and it is decelerated by the divergence of the horizontal Reynolds stress.

The mean zonal momentum balance from the computational solution is shown in Fig. 5.17. In the upper layer the eastward surface wind stress is balanced primarily by eddy isopycnal form stress (i.e., associated with the first interior interface between layers, $-\mathcal{D}_{1.5}$). The Reynolds stress, $\mathcal{R}$, divergence redistributes the zonal momentum in y, increasing the eastward momentum in the jet core and decreasing it at the jet edges (also known as negative eddy viscosity since the Reynolds stress is an up-gradient momentum flux relative to the mean horizontal shear). However, $\mathcal{R}$ cannot have any integrated effect since

$$\int_0^{L_y} dy \frac{\partial \mathcal{R}}{\partial y} = \mathcal{R}\Big|_0^{L_y} = 0 \tag{5.89}$$

due to the meridional boundary conditions (cf., Section 5.4). In the bottom layer eastward momentum is transmitted downward by the isopycnal form stress (i.e., $+\mathcal{D}_{2.5}$), and it is balanced almost entirely by the topographic and turbulent bottom stress because the abyssal $\mathcal{R}_N$ is quite weak. The shape for $\mathcal{R}(y)$ can be interpreted either in terms of radiating Rossby waves (Section 5.4) or as a property of the linearly unstable eigenmodes for $\bar{u}_n(y)$ (not shown here; cf., Section 3.3.3 for barotropic eigenmodes). When $L_y, L_\tau \gg L_\beta \gg R$, multiple jet cores can occur through the up-gradient fluxes by $\mathcal{R}$, each with a meridional scale near L_β. The scale relation, $L_y \gg L_\beta$, is only marginally satisfied for the westerly winds, but it is more likely true for the ACC, and some observational evidence indicates persistent multiple jet cores there.

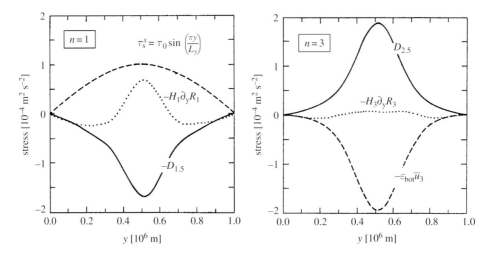

Fig. 5.17. Terms in the mean zonal momentum budget (5.82) in the upper-($n = 1$) and lower-most ($n = N$) layers for a quasigeostrophic zonal-jet solution with $N = 3$ (adapted from McWilliams & Chow, 1981).

5.3.4 Potential vorticity homogenization

From (5.27), (5.84), and (5.88), the mean zonal momentum balance (5.82) can be rewritten more concisely as

$$\frac{\partial \overline{u}_n}{\partial t}[=0] = \overline{v'_n q'_{\text{QG},n}} + \hat{\mathbf{x}} \cdot \overline{\mathbf{F}}_n, \qquad (5.90)$$

after doing zonal integrations by parts. This shows that the eddy–mean interaction for a baroclinic zonal-channel flow is entirely captured by the meridional eddy potential vorticity flux that combines the Reynolds stress and isopycnal form stress divergences:

$$\begin{aligned}\overline{v'_n q'_{\text{QG},n}} &= -\frac{\mathcal{D}_{1.5}}{H_1} - \frac{\partial \mathcal{R}_1}{\partial y}, \quad n = 1 \\[2mm]
&= \left(\frac{\mathcal{D}_{n-0.5} - \mathcal{D}_{n+0.5}}{H_n}\right) - \frac{\partial \mathcal{R}_n}{\partial y}, \quad 2 \le n \le N-1 \\[2mm]
&= \left(\frac{\mathcal{D}_{N-0.5} - \mathcal{D}_{\text{bot}}}{H_N}\right) - \frac{\partial \mathcal{R}_N}{\partial y}, \quad n = N. \end{aligned} \qquad (5.91)$$

In the vertical interior where $\overline{\mathbf{F}}_n$ is small, (5.90) indicates that $\overline{v'_n q'_{\text{QG},n}}$ is also small. Since q_{QG} is approximately conserved following parcels in (5.26), a fluctuating Lagrangian meridional parcel displacement, $r^{y'}$, generates a potential vorticity fluctuation,

$$q'_{\text{QG}} \approx -r^{y'}\frac{d\overline{q}_{\text{QG}}}{dy}, \qquad (5.92)$$

since potential vorticity is approximately conserved along trajectories (cf., Section 3.5). For non-zero $r^{y'}$, due to non-zero v', the required smallness of the eddy potential vorticity flux can be accomplished if q'_{QG} is small as a consequence of $d_y\overline{q}_{\text{QG}}$ being small. This is an explanation for the homogenized structure for the mid-depth potential vorticity profile, $q_{\text{QG},2}(y)$, seen in the second-row plots in Fig. 5.14. Furthermore, the variance for $q'_{\text{QG},2}$ (not shown) is also small even though the variances of other interior quantities are not small (Fig. 5.15).

Any other material tracer, τ, that is without either significant interior source or diffusion terms, $\mathcal{S}^{(\tau)}$ in (2.8), or boundary fluxes that maintain a mean gradient, $\overline{\tau}(y)$, will be similarly homogenized by eddy mixing in a statistical equilibrium state.

5.3.5 Meridional overturning circulation and mass balance

The relation (5.22), which expresses the movement of the interfaces as material surfaces, is single-valued in w at each interface because of the quasigeostrophic approximation (Section 5.1.2). In combination with the Ekman pumping at the interior edges of the embedded turbulent boundary sub-layers (Sections 5.3.1 and 6.1), w is a vertically continuous, piecewise linear function of depth within each layer. The time and zonal mean vertical velocity at the interior interfaces is

$$\overline{w}_{n+0.5} = \frac{\partial}{\partial y}\overline{v_{g,n+0.5}\eta_{n+0.5}} = -\frac{1}{f_0}\frac{\partial}{\partial y}\mathcal{D}_{n+0.5} \qquad (5.93)$$

for $1 \leq n \leq N-1$ (i.e., $\overline{w}$ is forced by the isopycnal form stress in the interior). The vertical velocities at the vertical boundaries are determined from the kinematic conditions. At the rigid lid (Section 2.2.3), $\overline{w} = 0$, and at the bottom,

$$\overline{w} = \overline{\mathbf{u}_N \cdot \nabla B} \approx \frac{\partial}{\partial y}\overline{v_{g,N}B} = -\frac{1}{f_0}\frac{\partial}{\partial y}\mathcal{D}_{\text{bot}},$$

from (5.86). Substituting the mean vertical velocity into the mean continuity relation (5.81) and integrating in y yields

$$\overline{v}_{a,1} = -\frac{\mathcal{D}_{1.5}}{f_0 H_1}$$

$$\overline{v}_{a,n} = \frac{\mathcal{D}_{n-0.5} - \mathcal{D}_{n+0.5}}{f_0 H_n}, \quad 2 \leq n \leq N-1$$

$$\overline{v}_{a,N} = \frac{\mathcal{D}_{N-0.5} - \mathcal{D}_{\text{bot}}}{f_0 H_N}. \qquad (5.94)$$

From the structure of the $\mathcal{D}_{n+0.5}(y)$ in Fig. 5.17, the *meridional overturning circulation* can be deduced. Because $\mathcal{D}(y)$ has a positive extremum at the jet center, (5.93) implies that $\overline{w}(y)$ is upward on the southern side of the jet and downward on the northern side. Mass conservation for the meridional overturning circulation is closed in the surface layer with a strong northward flow. In (5.94) this ageostrophic flow, $\overline{v}_{a,1} > 0$, is related to the downward isopycnal form stress, but in the zonal momentum balance for the surface layer (5.82) combined with (5.83) and (5.99), it is closely tied to the eastward surface stress as an *Ekman transport* (Section 6.1). Depending upon whether $\mathcal{D}(z)$ decreases or increases with depth, (5.94) implies that $\overline{v}_a$ is southward or northward in the interior. In the particular solution in Fig. 5.17, $\mathcal{D}$ weakly increases between interfaces $n+0.5 = 1.5$ and 2.5 because $\mathcal{R}_n$ decreases with depth in the middle of the jet. So $\overline{v}_{a,2}$ is weakly northward in the jet center. Because $\mathcal{D}_{N-1} > 0$, the bottom layer flow is southward, $\overline{v}_{a,N} < 0$. Furthermore, since the bottom-layer zonal flow is eastward, $\overline{u}_N > 0$, the associated bottom stress in (5.94) provides an augmentation to the southward $\overline{v}_{a,N}$ (NB this contribution is called the bottom Ekman transport; Section 6.1). Collectively, this structure accounts for the clockwise Deacon Cell depicted in Figs. 5.13 and 5.14.

In a layered model the pointwise continuity equation is embodied in the layer thickness equation (5.18) that also embodies the parcel conservation of density. Its time and zonal mean reduces to

$$\frac{\partial}{\partial y}\overline{h_n v_n} = 0 \implies \overline{h_n v_n} = 0, \tag{5.95}$$

using a boundary condition for no flux at some (remote) latitude to determine the meridional integration constant. In equilibrium there is no meridional mass flux within each isopycnal layer in an adiabatic fluid because the layer boundaries (bottom, interfaces, and lid) are material surfaces. This relation can be rewritten as

$$\overline{h_n v_n} = H_n \overline{v}_{a,n} + \overline{(h_n - H_n)v_{g,n}} = 0 \tag{5.96}$$

(cf., (5.83)). There is an exact cancelation between the mean advective mass flux (the first term) and the eddy-induced mass transport (the second term) within each isopycnal layer. The same conclusion about cancelation between the mean and eddy transports could be drawn for any non-diffusing tracer that does not cross the material interfaces.

Re-expressing the eddy mass flux in terms of a meridional *eddy-induced transport velocity* or *bolus velocity* defined by

$$V_n^* = \frac{1}{H_n}\overline{(h_n - H_n)v_{g,n}}, \tag{5.97}$$

the cancelation relation (5.96) becomes simply

$$\overline{v}_{a,n} = -V_n^*.$$

There is a companion vertical component to the eddy-induced velocity, $W_{n+0.5}^*$, that satisfies a continuity equation with the horizontal component, analogous to the 2D mean continuity balance (5.81). In a zonally symmetric channel flow, the eddy-induced velocity is 2D, as is its continuity balance:

$$\frac{\partial V_n^*}{\partial y} + \frac{1}{H_n}\left(W_{n-0.5}^* - W_{n+0.5}^*\right) = 0. \tag{5.98}$$

$\mathbf{U}^* = (0, V^*, W^*)$ has zero normal flow at the domain boundaries (e.g., $W_{0.5}^* = 0$ at the top surface). Together these components of the eddy-induced meridional overturning circulation exactly cancel the Eulerian mean Deacon Cell circulation, $(0, \overline{v}_a, \overline{w})$. One can interpret $\mathbf{U}^*$ as a Lagrangian mean circulation induced by the eddies that themselves have a zero Eulerian mean velocity. It is therefore like a Stokes drift (Section 4.5), but one caused by the mesoscale eddy velocity field rather than the surface or inertia-gravity waves. The mean fields for both mass and other material concentrations move with (i.e., are advected by) the sum of the Eulerian mean and eddy-induced Lagrangian mean velocities. Here the fact that their sum is zero in the meridional plane is due to the adiabatic assumption.

Expressing h in terms of the interface displacements, η, from (5.19) and $\mathcal{D}$ from (5.87), the mass balance (5.96) can be rewritten as

$$\overline{(h_n - H_n)\, v_{g,n}} = -H_n \bar{v}_{a,n}$$

$$= \frac{1}{f_0} \mathcal{D}_{1.5}, \quad n = 1$$

$$= -\frac{1}{f_0} \left(\mathcal{D}_{n-0.5} - \mathcal{D}_{n+0.5} \right), \quad 2 \le n \le N-1$$

$$= -\frac{1}{f_0} \left(\mathcal{D}_{N-0.5} - \mathcal{D}_{\text{bot}} \right), \quad n = N. \qquad (5.99)$$

This demonstrates an equivalence between the vertical isopycnal form stress divergence and the lateral eddy mass flux within an isopycnal layer.

5.3.6 Meridional heat balance

The buoyancy field, b, is proportional to η in (5.23). If the buoyancy is controlled by the temperature, T (e.g., as in the simple equation of state used here, $b = \alpha g T$), then the interfacial temperature fluctuation is defined by

$$T_{n+0.5} = -\frac{2 g'_{n+0.5}}{\alpha g (H_n + H_{n+1})} \eta_{n+0.5}. \qquad (5.100)$$

With this definition the meridional eddy heat flux is equivalent to the interfacial form stress (5.87), hence layer mass flux (5.99), by the following relation:

$$\overline{vT}_{n+0.5} = \frac{f_0 \mathcal{N}^2_{n+0.5}}{\alpha g} \mathcal{D}_{n+0.5}. \qquad (5.101)$$

The mean buoyancy frequency is defined by

$$\mathcal{N}^2_{n+0.5} = \frac{2 g'_{n+0.5}}{H_n + H_{n+1}}$$

analogous to (4.17). Since $\mathcal{D} > 0$ in the jet (Fig. 5.17), $\overline{vT} < 0$; i.e., the eddy heat flux is poleward in the ACC (cf., Section 5.2.3). The profile for $\overline{T}_{n+0.5}(y)$ (Fig. 5.14) indicates that this is a down-gradient eddy heat flux associated with release of mean available potential energy. These behaviors are hallmarks of baroclinic instability (Section 5.2).

The equilibrium heat balance at the layer interfaces is obtained by a reinterpretation of (5.93), replacing η by T from (5.100):

$$\frac{\partial}{\partial t}\overline{T}_{n+0.5}\,[=0] \;=\; -\frac{\partial}{\partial y}[\overline{vT}_{n+0.5}] - \overline{w}_{n+0.5}\,\partial_z\overline{T}_{n+0.5}. \qquad (5.102)$$

The background vertical temperature gradient, $\partial_z\overline{T}_{n+0.5,\,z} = \mathcal{N}^2_{n+0.5}/\alpha g$, is the mean stratification expressed in terms of temperature. Thus, the horizontal eddy heat-flux divergence is balanced by the mean vertical advection of the background temperature stratification in the equilibrium state.

5.3.7 Maintenance of the general circulation

In summary, the eddy fluxes for momentum, mass, and heat play essential roles in the equilibrium dynamical balances for the jet. In particular, $\mathcal{D}$ is the most important eddy flux, accomplishing the essential transport to balance the mean forcing. For the ACC the mean forcing is a surface stress, and $\mathcal{D}$ is most relevantly identified as the interfacial form stress that transfers the surface stress downward to push against the bottom (cf. (5.85)). For the atmospheric westerly winds, the mean forcing is the differential heating with latitude, and $\mathcal{D}$ plays the necessary role as the balancing poleward heat flux. Of course, both roles for $\mathcal{D}$ are played simultaneously in each case. The outcome in each case is an upward-intensified, meridionally sheared zonal mean flow, with associated sloping isopycnal and isothermal surfaces in thermal wind balance. It is also true that the horizontal Reynolds stress, $\mathcal{R}$, contributes to the zonal mean momentum balance and thereby influences the shape of $\overline{u}_n(y)$ and its geostrophically balancing geopotential and buoyancy fields, most importantly by sharpening the core jet profile. But $\mathcal{R}$ does not provide the essential equilibrating balance to the overall forcing (i.e., in the meridional integral of (5.82)) in the absence of meridional boundary stresses (cf. (5.89) and Section 5.4).

Much of the preceding dynamical analysis is a picture drawn first in the 1950s and 1960s to describe the maintenance of the atmospheric Jet Stream (e.g., Lorenz, 1967). Nevertheless, for many years afterward it remained a serious challenge to obtain computational solutions that exhibit this behavior. This GFD problem has such central importance, however, that its interpretation continues to be further refined. For example, it has recently become a common practice to diagnose the eddy effects in terms of the *Eliassen–Palm flux* defined by

$$\mathbf{E} = \overline{u'v'}\hat{\mathbf{y}} + f_0\overline{\eta'v'}\hat{\mathbf{z}} = \mathcal{R}\hat{\mathbf{y}} - \mathcal{D}\hat{\mathbf{z}}. \qquad (5.103)$$

(NB, $\mathbf{E}$ has a 3D generalization beyond the zonally symmetric channel flow considered here.) The ingredients of $\mathbf{E}$ are the eddy Reynolds stress, $\mathcal{R}$, and isopycnal form stress, $\mathcal{D}$. The mean zonal acceleration by the eddy fluxes in (5.90) is re-expressable as minus the divergence of the Eliassen–Palm flux, i.e.,

$$\overline{v'q'_{QG}} = -\boldsymbol{\nabla}\cdot\boldsymbol{E} = -\partial_y\mathcal{R} + \partial_z\mathcal{D},$$

with all the associated dynamical roles in the maintenance of the turbulent equilibrium jet that have been discussed throughout this section. (An analogous perspective for wind-driven oceanic gyres is in Section 6.2.)

The principal utility of a general circulation model – whether for the atmosphere, the ocean, or their coupled determination of climate – is in mediating the competition among external forcing, eddy fluxes, and non-conservative processes with as much geographical realism as is computationally feasible.

5.4 Rectification by Rossby-wave radiation

A mechanistic interpretation for the shape of $\mathcal{R}_n(y)$ in Fig. 5.17 can be made in terms of the eddy–mean interaction associated with Rossby waves radiating meridionally away from a source in the core region for the mean jet and dissipating after propagating some distance away from the core. For simplicity this analysis will be made with a barotropic model (cf., Section 3.4), since barotropic, shallow-water, and baroclinic Rossby waves are all essentially similar in their dynamics. The process of generating a mean circulation from transiently forced fluctuating currents is called *rectification*. In coastal oceans tidal rectification is common.

A non-conservative, barotropic, potential vorticity equation on the β-plane is

$$\frac{Dq}{Dt} = \mathcal{F}' - r\nabla^2\psi$$

$$q = \nabla^2\psi + \beta y$$

$$\frac{D}{Dt} = \frac{\partial}{\partial t} + \hat{\mathbf{z}}\cdot\boldsymbol{\nabla}\psi\times\boldsymbol{\nabla} \qquad\qquad (5.104)$$

(cf., (3.27)). For the purpose of illustrating rectification behavior, $\mathcal{F}'$ is a transient forcing term with zero time mean (e.g., caused by Ekman pumping from fluctuating winds), and r is a damping coefficient (e.g., Ekman drag; cf., (5.80) and (6.53) with $r = \epsilon_{\text{bot}}/H$). For specificity choose

$$\mathcal{F}' = F_*(x, y)\sin[\omega t],$$

with a localized F_* that is non-zero only in a central region in y (Fig. 5.18).

Rossby waves with frequency ω will be excited and propagate away from the source region. Their dispersion relation is

$$\omega = -\frac{\beta k}{k^2 + \ell^2}, \qquad\qquad (5.105)$$

with (k, ℓ) the horizontal wavenumber vector. The associated meridional phase and group speeds are

$$c_p^y = \omega/\ell = -\frac{\beta k}{\ell(k^2 + \ell^2)}$$

$$c_g^y = \frac{\partial \omega}{\partial \ell} = \frac{2\beta k \ell}{(k^2 + \ell^2)^2} \tag{5.106}$$

(Section 4.7). To the north of the source region, the group speed must be positive for outward energy radiation. Since without loss of generality $k > 0$, the northern waves must have $\ell > 0$. This implies $c_p^{(y)} < 0$ and a NW–SE alignment of the constant-phase lines, hence $\overline{u'v'} < 0$ since motion is parallel to the constant-phase lines. In the south the constant-phase lines have a NE–SW alignment, and $\overline{u'v'} > 0$. This leads to the $\overline{u'v'}(y)$ profile in Fig. 5.18. Note the decay as $|y| \to \infty$, due to damping by r. In the vicinity of the source region the flow can be complicated, depending upon the form of F_*, and here the far-field relations are connected smoothly across it without too much concern about local details.

   This Reynolds stress enters in the time-mean, zonal momentum balance consistent with (5.104):

$$r\bar{u} = -\frac{\partial}{\partial y}\left(\overline{u'v'}\right) \tag{5.107}$$

since $\bar{F} = 0$ (cf., Section 3.4). The mean zonal flow generated by wave rectification has the pattern sketched in Fig. 5.18, eastward in the vicinity of the source and westward to the north and south. This a simple model for the known behavior of eastward acceleration by the eddy horizontal momentum flux in a baroclinically unstable eastward jet (e.g., in the Jet Stream and ACC; Section 5.3.3), where the eddy generation process by baroclinic instability has been replaced heuristically by the transient forcing $\mathcal{F}'$. The mean flow profile in Fig. 5.18 is proportional to $-\partial_y \mathcal{R}$, and it has a shape very much like the one in Fig. 5.17. Note that this rectification process does not act like an eddy diffusion process in the generation region since $\overline{u'v'}$ generally has the same sign as $\bar{u}_y$ (and here it could, misleadingly, be called a negative eddy-viscosity process), although these quantities do have opposite signs in the far-field where the waves are being dissipated. So the rectification is not behaving like eddy mixing in the source region, in contrast to the barotropic instability problems discussed in Sections 3.3 and 3.4. The eddy process here is highly non-local, with the eddy generation site (within the jet) distant from the dissipation site (outside the jet). Since

$$\int_{-\infty}^{\infty} \bar{u}(y)\,dy = 0 \tag{5.108}$$

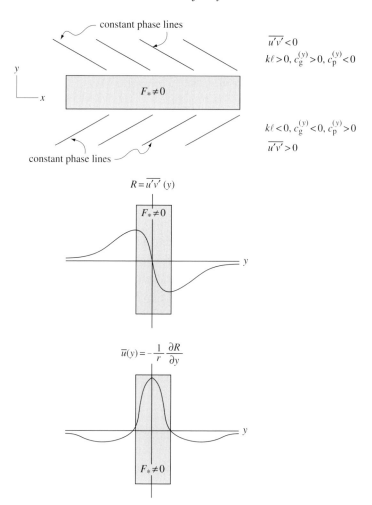

Fig. 5.18. Sketch of radiating Rossby waves from a zonal strip of transient forcing, $F_*(x, y, t)$ (shaded area); the pattern of Rossby-wave crests and troughs (i.e., lines of constant wave phase) consistent with the meridional group velocity, c_g^y, oriented away from the forcing strip (top); the resulting Reynolds stress, $\mathcal{R}(y)$ (middle); and the rectified mean zonal flow, $\bar{u}(y)$ (bottom). By the dispersion relation (5.105), the signs for $k\ell$ and c_p^y on either side of the forcing region are consequences of outward energy propagation.

from (5.107), the rectification process can be viewed as a conservative redistribution of the ambient zonal-mean zonal momentum, initially zero everywhere, through wave *radiation stresses*.

There are many other important examples of *non-local transport* of momentum by waves in nature. The momentum is taken away from where the waves are generated and deposited where they are dissipated. For example, this happens for *internal gravity lee waves* generated by a persistent flow (even by tides; Fig. 4.2) over a bottom topography on which they exert a mean form stress. The gravity lee waves propagate upward away from the solid boundary with a dominant wavenumber vector, $\mathbf{k}_*$, determined from their dispersion relation and the mean wind speed in order to be stationary relative to solid Earth. The waves finally break and dissipate mostly at *critical layers* (i.e., where $c_p(k_*) = \bar{u}(z)$), and the associated Reynolds stress divergence, $-\partial_z \overline{\mathbf{u}'w'}$, acts to retard the mean flow aloft. This process is an important influence on the strength of the tropopause Jet Stream, as well as mean zonal flows at higher altitudes. Perhaps it may be similarly important for the ACC as well, but the present observational data do not allow a meaningful test of this hypothesis.

6

Boundary-layer and wind-gyre dynamics

Boundary layers arise in many situations in fluid dynamics. They occur where there is an incompatibility between the interior dynamics and the boundary conditions, and a relatively thin transition layer develops with its own distinctive dynamics in order to resolve the incompatibility. For example, non-zero fluxes of momentum, tracers, or buoyancy across a fluid boundary almost always instigate an adjacent boundary layer with large normal gradients of the fluid properties. Boundary-layer motions typically have smaller spatial scales than the dominant interior flows. If their Re value is large, they have stronger fluctuations (i.e., eddy kinetic energy) than the interior because they almost always are turbulent. Alternatively, in a laminar flow with a smaller Re value, but still with interior dynamical balances that are nearly conservative, boundary layers develop where non-conservative viscous or diffusive effects are significant because the boundary-normal spatial gradients are larger than in the interior.

In this chapter two different types of boundary layers are examined. The first type is a *planetary boundary layer* that occurs near the solid surface at the bottoms of the atmosphere and ocean and on either side of the ocean–atmosphere interface. The instigating vertical boundary fluxes are a momentum flux – the drag of a faster moving fluid against slower (or stationary) material at the boundary – or a buoyancy flux – heat and water exchanges across the boundary. The second type is a *lateral boundary layer* that occurs, most importantly, at the western side of an *oceanic wind gyre* in an extra-tropical basin with solid boundaries in the zonal direction. It occurs in order to satisfy the constraint of zonally integrated mass conservation (i.e., zero total meridional transport in steady state) that the interior meridional currents by themselves do not.

6.1 Planetary boundary layer

The planetary boundary layer is a region of strong, 3D, nearly isotropic turbulence associated with motions of relatively small scale (1–10^3 m) that, nevertheless, are

186

often importantly influenced by Earth's rotation. Planetary boundary layers are found near all solid–surface, air–sea, air–ice, and ice–sea boundaries. The primary source of the turbulence is the instability of the $\overline{\rho}(z)$ and $\overline{\mathbf{u}}(z)$ profiles that develop strong vertical gradients in response to the boundary fluxes. For example, either a negative buoyancy flux (e.g., cooling) at the top of a fluid layer or a positive buoyancy flux at the bottom generates a gravitationally unstable density profile and induces convective turbulence (cf. Section 2.3.3). Similarly, a boundary stress caused by drag on the adjacent flow generates a strongly sheared, unstable velocity profile, inducing shear turbulence (cf. Section 3.3.3). In either case the strong turbulence leads to an efficient buoyancy and momentum mixing that has the effect of reducing the gradients in the near-boundary profiles. This sometimes happens to such a high degree that the planetary boundary layer is also called a *mixed layer*, especially with respect to the weakness in material tracer gradients.

A typical vertical thickness, h, for the planetary boundary layer is 50 m in the ocean and 500 m in the atmosphere, although wide ranges of h occur even on a daily or hourly basis (Fig. 6.1) as well as climatologically. The largest h values occur for a strongly destabilizing boundary buoyancy flux, instigating convective turbulence, where h can penetrate through most or all of either the ocean or the troposphere e.g. deep subpolar oceanic convection in the Labrador and Greenland Seas or deep tropical atmospheric convection above the Western Pacific Warm Pool. More often convective boundary layers do not penetrate throughout the fluid because their depth is limited by stable stratification in the interior (e.g., (6.1)), which is sometimes called a *capping inversion* or *inversion layer* in the atmosphere or a pycnocline in the ocean.

6.1.1 Boundary-layer approximations

The simplest example of a shear planetary boundary layer is a uniform-density fluid that is generated in response to the stress (i.e., momentum flux through the boundary) on an underlying flat surface at $z = 0$. The incompressible, rotating, momentum, and continuity equations with $\rho = \rho_0$ are

$$\frac{Du}{Dt} - fv = -\frac{\partial \phi}{\partial x} + F^x$$

$$\frac{Dv}{Dt} + fu = -\frac{\partial \phi}{\partial y} + F^y$$

$$\frac{Dw}{Dt} + g = -\frac{\partial \phi}{\partial z} + F^z$$

$$\frac{\partial u}{\partial x} + \frac{\partial v}{\partial y} + \frac{\partial w}{\partial z} = 0. \tag{6.1}$$

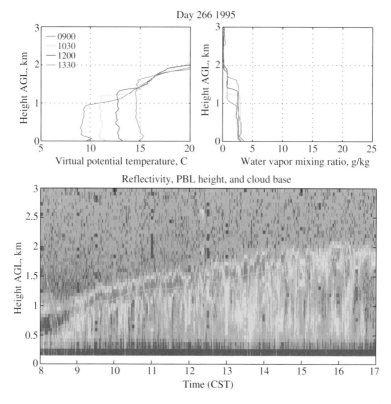

Fig. 6.1. Color plate 5.   Example of reflectivities (bottom) observed in the cloud-free convective boundary layer in central Illinois on 23 September 1995: (top left) virtual temperature profiles and (top right) vertical profiles of water vapor mixing ratio. Note the progressive deepening of the layer through the middle of the day as the ground warms (Angevine *et al.*, 1998).

This partial differential equation system has solutions with both turbulent fluc-tuations and a mean velocity component, where the mean is distinguished by an average over the fluctuations. So the planetary boundary layer is yet another geophysically important example of eddy–mean interaction.

Often, especially from a large-scale perspective, the mean boundary-layer flow and tracer profiles are the quantities of primary interest, and the turbulence is viewed as a distracting complexity, interesting only as a necessary ingredient for determining the mean velocity profile. From this perspective the averaged equations that express the *mean-field balances* are the most important ones. In the context of a general circulation model, for example, the mean-field balances are part of the model formulation with an appropriate parameterization for the averaged transport effects by the turbulent eddies. In the shear planetary boundary layer, the transport is expressed as the averaged *eddy momentum flux*, i.e., the Reynolds stress (Section 3.4).

To derive the mean-field balances for (6.1), all fields are decomposed into mean and fluctuating components,

$$u = \bar{u} + u', \quad \text{etc.} \tag{6.2}$$

For a boundary layer in the z direction, the overbar denotes an average in x, y, t over the scales of the fluctuations, so that, e.g.,

$$\overline{u'} = 0. \tag{6.3}$$

This technique presumes a degree of statistical symmetry in these averaging coordinates, at least on the typical space and time scales of the fluctuations. No average is taken in the z direction since both fluctuation and mean variables will have strong z gradients and not be translationally symmetric in z. Alternatively, the average may be viewed as over an *ensemble* of many planetary boundary layer realizations with the same mean stress and different initial conditions for the fluctuations, counting on the sensitive dependence of the solutions to (6.1) spanning the range of possible fluctuation behaviors. If there is a separation of space and/or time scales between the mean and fluctuating components (e.g., as assumed in Section 3.5), and if there is a meaningful typical statistical equilibrium state for all members of the ensemble, then it is usually presumed that the symmetry-coordinate and ensemble averages give equivalent answers. This presumption is called *ergodicity*. For highly turbulent flows in GFD, the ergodicity assumption is usually valid.

The momentum advection term can be rewritten as a momentum flux divergence, namely,

$$(\mathbf{u} \cdot \boldsymbol{\nabla})\mathbf{u} = \boldsymbol{\nabla} \cdot (\mathbf{uu}), \tag{6.4}$$

with a 3D vector notation. This equivalence is valid since the difference between the two sides of the equation, $\mathbf{u}(\boldsymbol{\nabla} \cdot \mathbf{u})$, vanishes by the incompressible continuity relation. An average of the quadratic momentum flux yields two types of contributions,

$$\boldsymbol{\nabla} \cdot (\overline{\mathbf{uu}}) = \boldsymbol{\nabla} \cdot (\bar{\mathbf{u}}\,\bar{\mathbf{u}}) + \boldsymbol{\nabla} \cdot (\overline{\mathbf{u}'\mathbf{u}'}), \tag{6.5}$$

since terms that are linear in the fluctuations vanish by (6.3) while those that are quadratic do not. Equation (6.5) is an expression for the divergence of the mean momentum flux. The averaged advective flux thus has contributions both from the mean motions (the first term, the mean momentum flux) and the fluctuations (the second term, the eddy momentum flux or Reynolds stress).

Insert (6.2) into (6.1) and take the average. The result is

$$\frac{\partial \overline{u}}{\partial t} + \nabla \cdot (\overline{\mathbf{u}}\ \overline{u}) - f\overline{v} = -\frac{\partial \overline{\phi}}{\partial x} - \nabla \cdot (\overline{\mathbf{u}'u'}) + \overline{F^x}$$

$$\frac{\partial \overline{v}}{\partial t} + \nabla \cdot (\overline{\mathbf{u}}\ \overline{v}) + f\overline{u} = -\frac{\partial \overline{\phi}}{\partial y} - \nabla \cdot (\overline{\mathbf{u}'v'}) + \overline{F^y}$$

$$\frac{\partial \overline{w}}{\partial t} + \nabla \cdot (\overline{\mathbf{u}}\ \overline{w}) + g = -\frac{\partial \overline{\phi}}{\partial z} - \nabla \cdot (\overline{\mathbf{u}'w'}) + \overline{F^z}$$

$$\frac{\partial \overline{u}}{\partial x} + \frac{\partial \overline{v}}{\partial y} + \frac{\partial \overline{w}}{\partial z} = 0. \tag{6.6}$$

The mean linear terms simply match those in the unaveraged equations (6.1), and the quadratic terms additionally contain the Reynolds stress.

Now make assumptions about the dynamical balances in (6.6) that comprise the *boundary-layer approximation*. Assume that (x, y, t) derivatives are small when applied to mean fields (including the eddy flux, which is a "mean" field since it is an averaged quantity) compared to vertical derivatives. The vertical scales of the turbulence and mean flow are both assumed to be on the scale of h, the boundary-layer thickness. Further assume that the mean and turbulent horizontal velocity magnitudes are comparable in size,

$$\overline{u}, \overline{v} \sim u', v'. \tag{6.7}$$

These assumptions imply that

$$\frac{\partial \overline{u}}{\partial x}, \frac{\partial \overline{u}}{\partial y}, \frac{\partial \overline{v}}{\partial x}, \frac{\partial \overline{v}}{\partial y} \ll \frac{\partial u'}{\partial x}, \frac{\partial u'}{\partial y}, \frac{\partial v'}{\partial x}, \frac{\partial v'}{\partial y}, \tag{6.8}$$

and

$$\overline{w} \ll w'. \tag{6.9}$$

These conclusions follow from the premises that both the mean and fluctuations satisfy a fully 3D continuity balance and that the flow is approximately horizontally *isotropic* (i.e., the x and y scales and u' and v' amplitudes are similar on average). If the horizontal scale of u', v' is comparable to h, the vertical scale, then the same consideration implies that w' is of comparable intensity to the horizontal velocities, and the turbulent motions are 3D isotropic. For dynamical consistency, the mean horizontal pressure gradient, $\nabla \overline{\phi}$, is assumed to be of the same size as the leading-order terms in the mean momentum balance, e.g., the Coriolis force, $f\overline{\mathbf{u}}_h$. In the boundary-layer approximation, the fluctuations are assumed to be statistically invariant in their horizontal and time dependences on the scales over which the mean flow varies. These approximations are called *homogeneity* (in the horizontal) and *stationarity* (in time), referring to the respective statistical properties of the fluctuations.

Next make a further approximation that the eddy flux divergences are much larger than the mean non-conservative force, $\overline{\mathbf{F}}$. Since

$$\mathbf{F} = \nu \nabla^2 \mathbf{u} \tag{6.10}$$

for Newtonian viscous diffusion (with ν the *viscosity*), then a scale estimate for the ratio of the eddy terms to the viscous diffusion terms in (6.6) is

$$Re = \frac{VL}{\nu}, \tag{6.11}$$

the Reynolds number previously defined in (2.5). It can be estimated for atmospheric/oceanic planetary boundary layers, respectively, with characteristic scales of $V \sim 10/0.1\,\mathrm{m\ s^{-1}}$, $L \sim h = 10^3/10^2\,\mathrm{m}$, and $\nu = 10^{-5}/10^{-6}\,\mathrm{m^2 s^{-1}}$, yielding the quite large values of $Re = 10^9/10^7$. Thus, the mean viscous diffusion, $\overline{\mathbf{F}}$, can be neglected in the mean-field balance (6.6).

Of course, $\mathbf{F'} = \nu \nabla^2 \mathbf{u'}$ cannot be neglected in (6.1) since a characteristic of turbulence is that the advective cascade of variance dynamically connects the large-scale fluctuations on the scale of h with small-scale fluctuations and molecular dissipation (cf., Sections 3.7 and 5.3). In the case of *three-dimensional turbulence* in general, and boundary-layer turbulence in particular, the fluctuation kinetic energy and enstrophy are both cascaded in the forward direction to the small, viscously controlled scales where it is dissipated – like the enstrophy cascade but unlike the energy cascade in 2D turbulence (Section 3.7).

The consequence of this boundary-layer approximation is a simplified form of the mean-field balances compared to (6.6), namely,

$$-f\overline{v} + \frac{\partial \overline{\phi}}{\partial x} + \frac{\partial}{\partial z}(\overline{w'u'}) = 0$$

$$f\overline{u} + \frac{\partial \overline{\phi}}{\partial y} + \frac{\partial}{\partial z}(\overline{w'v'}) = 0$$

$$\frac{\partial \overline{\phi}}{\partial z} + g + \frac{\partial}{\partial z}(\overline{w'^2}) = 0$$

$$\frac{\partial \overline{u}}{\partial x} + \frac{\partial \overline{v}}{\partial y} + \frac{\partial \overline{w}}{\partial z} = 0. \tag{6.12}$$

The equilibrium shear planetary boundary layer has a mean geostrophic, hydrostatic balance augmented by vertical eddy momentum flux divergences.

6.1.2 The shear boundary layer

For a shear planetary boundary layer next to a solid lower boundary, the following boundary conditions for the eddy fluxes are assumed:

$$\overline{w'u'} = -\frac{1}{\rho_0}\tau_{\mathrm{s}}^x, \quad \overline{w'v'} = -\frac{1}{\rho_0}\tau_{\mathrm{s}}^y, \quad \overline{w'^2} = 0, \quad z = 0, \tag{6.13}$$

and

$$\overline{w'u'}, \ \overline{w'v'}, \ \overline{w'^2} \to 0, \quad z \to \infty. \tag{6.14}$$

The mean boundary stress, τ_{s}, is conveyed toward the interior by the Reynolds stress that varies across the boundary layer, but decays away into the interior. Of course, the eddy flux cannot truly carry the stress very near the boundary since the velocity fluctuations, hence the Reynolds stress, must vanish there by the boundary condition of no slip. In reality a very thin *viscous sub-layer* lies next to the boundary, and within it the value of Re, expressed in terms of the local velocity and sub-layer thickness, is not large. In this sub-layer the important averaged momentum flux is the viscous one, $\nu\partial_z\overline{\mathbf{u}}$. So (6.13) represents a simplification by not resolving the viscous stress contribution in the sub-layer. Thus, the more general expression for the averaged vertical flux of horizontal momentum is

$$-\overline{w'\mathbf{u}_{\mathrm{h}}'} + \nu\frac{\partial\overline{\mathbf{u}}}{\partial z}.$$

Here τ_{s} is defined as the boundary stress exerted by the fluid on the underlying boundary or, equivalently, as the negative of the stress exerted by the solid boundary on the fluid.

Real boundary layers often have rather sharp vertical transitions in the fluctuation intensity and mean shear across the interior edge at $z \approx h$ as a consequence of the stable stratification there (e.g., a capping inversion). In (6.14) without stratification effects, the transition between the planetary boundary layer and interior regions is more gradual, and this interior edge has been identified with $z \to \infty$ for mathematical convenience.

Next decompose the solution of (6.12) into boundary-layer (with superscript b) and interior (with i) parts. The interior solutions satisfy geostrophic and hydrostatic balances since the eddy fluxes vanish by (6.14). Since the horizontal density gradients are zero here, the resulting thermal-wind balance (2.105) implies that the horizontal velocities are independent of depth,

$$\frac{\partial\overline{\mathbf{u}}_{\mathrm{h}}^{\mathrm{i}}}{\partial z} = 0. \tag{6.15}$$

This result is sometimes called the *Taylor–Proudman theorem*. The associated interior geopotential function and vertical velocity in the interior are therefore linear functions of z. So the structure of the mean flow solution is

$$\bar{u} = \bar{u}^{\mathrm{b}}(z) + \bar{u}^{\mathrm{i}}$$
$$\bar{v} = \bar{v}^{\mathrm{b}}(z) + \bar{v}^{\mathrm{i}}$$
$$\bar{w} = \bar{w}^{\mathrm{b}}(z) + \bar{w}^{\mathrm{i}}(z)$$
$$\bar{\phi} = x\overline{\mathcal{X}}^{\mathrm{i}} + y\overline{\mathcal{Y}}^{\mathrm{i}} + \phi_o - gz + \overline{\phi}^{\mathrm{b}}(z), \qquad (6.16)$$

making the z dependences explicit. $(\overline{\mathcal{X}}^{\mathrm{i}}, \overline{\mathcal{Y}}^{\mathrm{i}})$ is minus the mean horizontal pressure-gradient force, and ϕ_o is a reference constant for ϕ equal to its surface value. All of these mean-flow quantities can be viewed as having "slow" (x, y, t) variations on scales much larger than the scales of the boundary-layer fluctuations, although here this dependence is notationally suppressed.

The interior horizontal momentum balance is particularly simple:

$$f\bar{v}^{\mathrm{i}} \approx f\bar{v}^{\mathrm{i}}_{\mathrm{g}} = \overline{\mathcal{X}}^{\mathrm{i}}$$
$$f\bar{u}^{\mathrm{i}} \approx f\bar{u}^{\mathrm{i}}_{\mathrm{g}} = -\overline{\mathcal{Y}}^{\mathrm{i}}. \qquad (6.17)$$

The associated mean interior continuity balance implies that $\bar{w}^{\mathrm{i}}$ is trivial with this geostrophic flow if we neglect the spatial variation in f (i.e., f-plane approximation). A non-trivial balance for $\bar{w}^{\mathrm{i}}$ only occurs at a higher order of approximation, involving β and the horizontal ageostrophic interior flow, $\mathbf{u}^{\mathrm{i}}_{\mathrm{a}}$. The vertical integral of the continuity equation for the interior variables yields

$$\bar{w}^{\mathrm{i}} = \int^{z} \left(\partial_x \left[\frac{\overline{\mathcal{Y}}^{\mathrm{i}}}{f} \right] - \partial_y \left[\frac{\overline{\mathcal{X}}^{\mathrm{i}}}{f} \right] - \mathbf{V}_{\mathrm{h}} \cdot \bar{\mathbf{u}}^{\mathrm{i}}_{\mathrm{a}} \right) dz'$$
$$= \frac{\beta}{f^2} \overline{\mathcal{X}}^{\mathrm{i}} z - \int_0^z \mathbf{V}_{\mathrm{h}} \cdot \bar{\mathbf{u}}^{\mathrm{i}}_{\mathrm{a}} dz', \qquad (6.18)$$

with use of the surface boundary condition, $\bar{w}^{\mathrm{i}}(0) = 0$, in the second line. Since $\bar{\mathbf{u}}^{\mathrm{i}}_{\mathrm{a}}$ is undetermined at the level of approximation used in (6.12), the formula for $\bar{w}^{\mathrm{i}}$ cannot be explicitly evaluated at this point (and it is not the focus of this boundary-layer analysis).

Subtracting (6.17) from (6.12) yields the shear *boundary-layer problem*,

$$f\overline{v}^b(z) = \frac{\partial}{\partial z}(\overline{w'u'})$$

$$f\overline{u}^b(z) = -\frac{\partial}{\partial z}(\overline{w'v'})$$

$$\overline{\phi}^b(z) = -\overline{w'^2}$$

$$\frac{\partial \overline{w}^b}{\partial z} = -\frac{\partial \overline{u}^b}{\partial x} - \frac{\partial \overline{v}^b}{\partial y}, \qquad (6.19)$$

with the vertical boundary conditions,

$$\overline{\mathbf{u}}^b = -\overline{\mathbf{u}}^i, \quad \overline{w}^b = 0, \quad z = 0$$

$$\overline{u}^b, \overline{v}^b \to 0, \quad z \to \infty, \qquad (6.20)$$

appropriate for no-slip and no-normal flow at $z = 0$ and for vanishing horizontal boundary-layer velocities outside the layer. The third equation in (6.19) can be viewed as an auxiliary, diagnostic relation for the boundary-layer pressure correction to hydrostatic balance, since $\overline{\phi}^b$ does not influence the horizontal mean flow in these equations. Similarly, $\overline{w}^b$ is diagnostically determined from $\overline{\mathbf{u}}^b$ using the fourth equation in (6.19). So the central problem for the boundary layer is solving for $\overline{\mathbf{u}}_h^b$ using the first and second equations.

$\overline{w}$ differs from $(\overline{u}, \overline{v})$ in its interior and boundary-layer decomposition because $\overline{w}^b$ does not vanish as $z \to \infty$ and because the vertical boundary condition, $\overline{w}(0) = 0$, has, without loss of generality, been presumed to apply to each of $\overline{w}^b$ and $\overline{w}^i$ separately. The dynamical consistency of the latter presumption requires consistency with the boundary condition on $\overline{w}$ at the top of the fluid, $z = H$. From (6.18) this in turn is controlled by a consistent prescription for $\overline{\mathbf{u}}_a^i$, not explicitly considered here.

The quantity $\overline{w'\mathbf{u}'}(z)$ is called the vertical Reynolds stress because of its vertical flux direction for vector momentum (i.e., its first velocity component is w'). If it were known, then (6.19) and (6.20) could be solved to evaluate $\overline{u}^b$, $\overline{v}^b$, $\overline{w}_z^b$, and $\overline{\phi}^b$. However, the Reynolds stress is not known a priori, and at a fundamental dynamical level the eddy fluxes must be solved for simultaneously with the mean profiles; this requires looking beyond the mean-field balances (6.6). Seeking a way to avoid the full burden of the requirement of solving (6.1) completely is referred to as the *turbulence closure problem*, which if solved permits the consideration of (6.6) or (6.19) and (6.20) by themselves.

Even without specifying the closure, however, the horizontal momentum equations in (6.19) and (6.20) can be integrated across the boundary layer, eliminating the unknown Reynolds stress in place of the boundary stress using (6.13):

$$T^x = \int_0^\infty \overline{u}^b dz = -\frac{1}{\rho_0 f} \tau_s^y$$

$$T^y = \int_0^\infty \overline{v}^b dz = \frac{1}{\rho_0 f} \tau_s^x. \tag{6.21}$$

This says that the bottom boundary-layer, horizontal *transport* (i.e., depth-integrated velocity increment from the interior flow), $\mathbf{T}$, is 90° to the left (right) of the stress exerted by the fluid on the boundary in the northern (southern) hemisphere. The term transport is used both for depth-integrated horizontal velocity (i.e., a vertical-column area transport with units of $m^2 \, s^{-1}$) and for the normal component of horizontal velocity integrated over a vertical plane (i.e., a cross-section volume transport with units of $m^3 \, s^{-1}$, or when multiplied by ρ, a mass transport with units of $kg \, s^{-1}$).

The preceding boundary-layer analysis has neglected the horizontal derivatives of averaged quantities. The boundary-layer problem for $(\overline{u}^b, \overline{v}^b)$ in (6.19) and (6.20) and its associated transport (6.21) can be viewed as locally valid at each horizontal location. But by looking across different locations, the continuity equation in (6.19) can be integrated vertically with the kinematic condition in (6.20) to yield

$$\overline{w}^b(z \to \infty) = -\int_0^\infty \left(\frac{\partial \overline{u}^b}{\partial x} + \frac{\partial \overline{v}^b}{\partial y} \right) dz$$

$$\implies \overline{w}^b(\infty) = w_{ek,bot} = \hat{\mathbf{z}} \cdot \mathbf{\nabla} \times \left[\frac{\tau_s}{\rho_0 f} \right]. \tag{6.22}$$

Since $\overline{u}_h^b \to 0$ going into the interior, $\overline{w}^b$ approaches a value independent of height, which is how w_{ek} should be understood. Note that w_{ek}, like $\mathbf{T}$ in (6.21), depends only on the surface stress and is independent of the Reynolds stress profile, hence its closure.

Planetary boundary-layer flows that satisfy (6.19)–(6.22) are called *Ekman layers*. The vertical velocity that reaches into the interior, w_{ek} in (6.22), is called *Ekman pumping*, and it is caused by the horizontal gradients in τ_s and $f(y)$ on a spatial scale much larger than h (i.e., the boundary-layer approximation in Section 6.1.1).

The surface boundary conditions on $\overline{w'\mathbf{u}'}$ and $\overline{\mathbf{u}}^b$ in (6.13) and (6.20) are generally redundant for general $\overline{\mathbf{u}}^i$, although this viewpoint is based on the resolution of the closure problem alluded to above so that $\overline{w'\mathbf{u}'_h}$ and τ_s are mutually consistent.

- For the *bottom planetary boundary layers* in both the ocean and atmosphere, $\overline{\mathbf{u}}^i$ is locally viewed (i.e., from the perspective of the slowly varying (x, y, t) values) as determined from the interior dynamics independent of the details of the boundary-layer flow, and the planetary boundary-layer dynamics resolves the incompatibility between $\overline{\mathbf{u}}^i(z)$ and the surface no-slip boundary condition, $\overline{\mathbf{u}}(0) = 0$. This resolution determines the boundary-layer Reynolds stress profile, and the surface stress, τ_s, is diagnostically calculated using (6.20) (Section 6.1.4).
- In contrast, for the *oceanic surface planetary boundary layer* (with $z = 0$ the top surface and $z \rightarrow -\infty$ going into the oceanic interior), the overlying wind locally determines the surface stress, τ_s, and the surface oceanic velocity, $\overline{\mathbf{u}}^b(0) + \overline{\mathbf{u}}^i(0)$, is not constrained to be zero; in fact, the approximation is often made that the oceanic planetary boundary-layer dynamics are independent of the oceanic interior flow. Thus, (6.13) is the controlling boundary condition in the oceanic surface planetary boundary layer (Section 6.1.5).

(The more general perspective would be that the fluid dynamics of the interior and boundary layer, as well as the boundary stress, must all be determined together.)

For simplicity in this Ekman layer analysis, the assumptions $\rho = \rho_0$ and $\partial_z \overline{\mathbf{u}}^i = 0$ have been made. In general, the lower atmosphere and ocean are stratified, so these assumptions are not correct. However, they need not be precisely true to have planetary boundary-layer behavior that is very much like an Ekman layer, if the density variations and interior-flow baroclinicity (i.e., vertical shear) are weak enough near $z = 0$ on the scale $\delta z \sim h$ for both the turbulence and mean boundary-layer shear. Frequently the largest effect of stable stratification on a shear boundary layer is a compression of its vertical extent (i.e., reduction in h) by a strong pycnocline or inversion layer at its interior edge (cf. Fig. 6.1).

6.1.3 Eddy-viscosity closure

The preceding analysis is incomplete because the turbulence closure problem has not been resolved yet. One way to proceed is to adopt what is probably the most widely used *closure hypothesis* of *eddy viscosity*. It states that the Reynolds stress acts to transport momentum the same way that molecular viscosity does in (6.10), albeit with an enhanced eddy viscosity magnitude, $\nu_e \gg \nu$, whenever $Re \gg 1$ (Section 3.4). Specifically, in order to close (6.19) and (6.20), assume the following relation between the eddy momentum flux and the mean shear,

$$\overline{w'\mathbf{u}'} = -\nu_e \frac{\partial \overline{\mathbf{u}}}{\partial z}. \tag{6.23}$$

$\nu_{\mathrm{e}} > 0$ implies that the flux is acting in a down-gradient direction. Using the closure (6.23), the *laminar Ekman layer* equations are obtained from (6.19) and (6.20) for a boundary layer above a solid level surface:

$$f\overline{v} = -\nu_{\mathrm{e}}\frac{\partial^2 \overline{u}}{\partial z^2}$$

$$f\overline{u} = \nu_{\mathrm{e}}\frac{\partial^2 \overline{v}}{\partial z^2}$$

$$\overline{w} = -\int_0^z \left[\frac{\partial \overline{u}}{\partial x} + \frac{\partial \overline{v}}{\partial y}\right] dz'$$

$$\overline{\mathbf{u}}_{\mathrm{h}} = -\overline{\mathbf{u}}_{\mathrm{h}}^{\mathrm{i}}, \quad z = 0$$

$$\overline{\mathbf{u}}_{\mathrm{h}} \to 0, \quad z \to \infty. \tag{6.24}$$

For brevity the superscript on $\overline{\mathbf{u}}_{\mathrm{h}}^{\mathrm{b}}$ has now been deleted, and the accompanying diagnostic relation for $\overline{\phi}^{\mathrm{b}}$ is ignored.

This laminar boundary-layer problem is called a *parameterized planetary boundary layer* in the sense that there is no explicitly turbulent component in its solution, as long as ν_{e} is large enough that its corresponding eddy Reynolds number, $Re_{\mathrm{e}} = VL/\nu_{\mathrm{e}}$, is below some critical threshold value for instability of the mean boundary-layer velocity profile. (This reflects a general view that Re is a control parameter that regulates the transition from stable, laminar flow when Re is small, through transitional instabilities as Re increases, to fully developed turbulence when Re is large enough; cf., Section 5.3.1.) Nevertheless, the parameterized planetary boundary layer is implicitly a representation of (or model for) the intrinsically turbulent boundary-layer dynamics. The transport and Ekman pumping relations (6.21) and (6.22) are fully applicable to (6.24) since they do not depend on the Reynolds stress profile or closure choice.

It is quite feasible to use a closure relation like (6.23) in a general circulation model and implicitly solve for $\overline{\mathbf{u}}_{\mathrm{h}}^{\mathrm{b}}$ and $\overline{w}^{\mathrm{b}}$ as part of the time integration procedure for the model, and broadly speaking this approach is the common practice for incorporating planetary boundary-layer processes in general circulation models.

6.1.4 Bottom Ekman layer

Now solve the problem (6.24) analytically. The simplest way to do so is to define a complex horizontal velocity combination of the real velocity components,

$$U = \overline{u} + i\overline{v}. \tag{6.25}$$

In terms of U, (6.24) becomes

$$\nu_e \frac{\partial^2 U}{\partial z^2} = i f U$$

$$U(0) = -U^i$$

$$U(\infty) = 0. \tag{6.26}$$

This is a complex, second-order, ordinary differential equation boundary-value problem (rather than the equivalent, coupled pair of second-order equations, or a fourth-order system, for $(\bar{u}, \bar{v})$). This homogeneous problem has elemental solutions,

$$U \propto e^{kz} \quad \text{for} \quad k^2 = i\frac{f}{\nu_e},$$

or

$$k = (i\mathcal{S}_f)^{1/2} \sqrt{\frac{|f|}{\nu_e}}, \tag{6.27}$$

where $\mathcal{S}_f = f/|f|$ is equal to $+1$ in the northern hemisphere and -1 in the southern hemisphere. To satisfy $U(\infty) = 0$, the real part of k must be negative. This occurs for the k root,

$$(i\mathcal{S}_f)^{1/2} = -\frac{1+i\mathcal{S}_f}{\sqrt{2}}, \tag{6.28}$$

and the other root,

$$(i\mathcal{S}_f)^{1/2} = +\frac{1+i\mathcal{S}_f}{\sqrt{2}},$$

is excluded. The fact that k has an imaginary part implies an oscillation of U with z, in addition to the decay in z. Thus,

$$U(z) = -U^i e^{-\lambda(1+i\mathcal{S}_f)z}, \tag{6.29}$$

with a vertical decay rate, λ, that can be identified with the inverse boundary-layer depth by the relation,

$$h_{ek} = \lambda^{-1} = \sqrt{\frac{2\nu_e}{f}}. \tag{6.30}$$

This solution can be rewritten in terms of its real-valued velocity components from (6.25) as

$$\bar{u} = \text{Real}(U) = e^{-\lambda z}\left(-\bar{u}^i \cos[\lambda z] - \mathcal{S}_f \bar{v}^i \sin[\lambda z]\right)$$

$$\bar{v} = \text{Imag}(U) = e^{-\lambda z}\left(\mathcal{S}_f \bar{u}^i \sin[\lambda z] - \bar{v}^i \cos[\lambda z]\right). \tag{6.31}$$

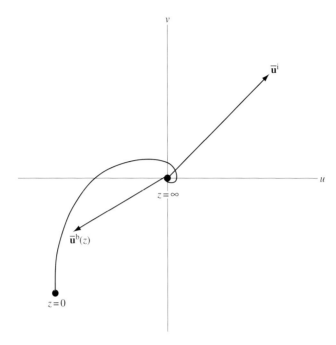

Fig. 6.2. Bottom Ekman layer hodograph in the northern hemisphere: $\overline{\mathbf{u}}^{\mathrm{i}}$ is the interior flow, and $\overline{\mathbf{u}}^{\mathrm{b}}(z)$ is the boundary-layer anomaly that brings the total flow to zero at the surface, $z = 0$.

The *Ekman spiral* is the curved plot of (6.31) that is evident on its *hodograph*, which is a plot of $\overline{\mathbf{u}}(z)$ in a (u, v) plane (Fig. 6.2). In the northern hemisphere the boundary layer velocity weakens and turns in a clockwise manner ascending from the surface, and the total velocity, $\mathbf{u}^{\mathrm{i}} + \mathbf{u}^{\mathrm{b}}$, increases from zero and also turns clockwise as it approaches the interior velocity. In the southern hemisphere the direction of the ascending Ekman spiral is counterclockwise. The spiral pattern of the hodograph is equivalent to the vertical decay and oscillation implied by k in (6.27) and (6.28).

Next evaluate the *surface stress* by differentiation of (6.31) using the eddy viscosity closure relation (6.23) evaluated at the bottom boundary:

$$\frac{1}{\rho_0}\tau_{\mathrm{s}}^x = \nu_{\mathrm{e}}\frac{\partial \overline{u}}{\partial z}(0) = \epsilon_{\mathrm{ek,bot}}\left(\overline{u}^{\mathrm{i}} - \mathcal{S}_f\overline{v}^{\mathrm{i}}\right)$$

$$\frac{1}{\rho_0}\tau_{\mathrm{s}}^y = \nu_{\mathrm{e}}\frac{\partial \overline{v}}{\partial z}(0) = \epsilon_{\mathrm{ek,bot}}\left(\mathcal{S}_f\overline{u}^{\mathrm{i}} + \overline{v}^{\mathrm{i}}\right), \tag{6.32}$$

with a bottom damping coefficient defined by

$$\epsilon_{\mathrm{ek,bot}} = \sqrt{\frac{|f|\nu_{\mathrm{e}}}{2}} = \frac{fh_{\mathrm{ek}}}{2} \tag{6.33}$$

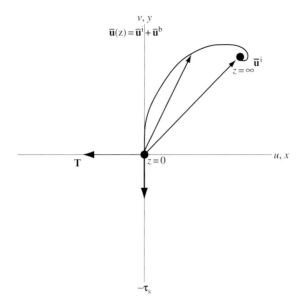

Fig. 6.3. Total velocity, $\bar{u}(z)$, boundary-layer transport, **T**, and boundary drag stress on the overlying fluid, τ_{s}, for a bottom Ekman layer in the northern hemisphere.

that has units of $\mathrm{m\,s^{-1}}$ (cf. (5.80) and Section 6.2.1). These relations indicate that the stress of the fluid on the boundary is rotated by $45°$ relative to the interior flow, and the rotation is to the left (right) in the northern (southern) hemisphere.

The transport (6.21) is evaluated by a direct vertical integration of (6.31) (with a result that necessarily must be consistent with the general relation (6.32)):

$$T^x = -\frac{\epsilon_{\mathrm{ek,bot}}}{f}\left(\mathcal{S}_f \bar{u}^{\mathrm{i}} + \bar{v}^{\mathrm{i}}\right) = -\frac{1}{\rho_0 f}\,\tau_{\mathrm{s}}^y$$

$$T^y = \frac{\epsilon_{\mathrm{ek,bot}}}{f}\left(\bar{u}^{\mathrm{i}} - \mathcal{S}_f \bar{v}^{\mathrm{i}}\right) = +\frac{1}{\rho_0 f}\,\tau_{\mathrm{s}}^x. \qquad (6.34)$$

In performing this integration, use was made of the definite integrals,

$$\int_0^\infty e^{-z}\cos[z]\,dz = \int_0^\infty e^{-z}\sin[z]\,dz = \frac{1}{2}. \qquad (6.35)$$

The boundary-layer transport is $135°$ to the left (right) of the interior flow, and the drag of the fluid on the boundary, τ_{s}, is oriented $45°$ to the left (right) in the northern (southern) hemisphere (Fig. 6.3). The Ekman pumping (6.22) is

$$\overline{w}^{\mathrm{b}}(z \to \infty) = \hat{\mathbf{z}}\cdot\boldsymbol{\nabla}\times\frac{\tau_{\mathrm{s}}}{\rho_o f}$$

$$= \frac{\partial}{\partial x}\left[\frac{\epsilon_{\mathrm{ek,bot}}}{f}(\mathcal{S}_f \bar{u}^{\mathrm{i}} + \bar{v}^{\mathrm{i}})\right] - \frac{\partial}{\partial y}\left[\frac{\epsilon_{\mathrm{ek,bot}}}{f}(\bar{u}^{\mathrm{i}} - \mathcal{S}_f \bar{v}^{\mathrm{i}})\right]$$

$$= \frac{\epsilon_{\text{ek,bot}}}{f} \left(\mathcal{S}_f \frac{\partial \overline{u}^{\text{i}}}{\partial x} + \frac{\partial \overline{v}^{\text{i}}}{\partial x} - \frac{\partial \overline{u}^{\text{i}}}{\partial y} + \mathcal{S}_f \frac{\partial \overline{v}^{\text{i}}}{\partial y} \right)$$

$$+ \frac{\beta \epsilon_{\text{ek,bot}}}{f^2} \left(\overline{u}^{\text{i}} - \mathcal{S}_f \overline{v}^{\text{i}} \right)$$

$$= \frac{\epsilon_{\text{ek,bot}}}{f} \overline{\zeta}^{\text{i}} + \frac{\beta \epsilon_{\text{ek,bot}}}{f^2} \left(\overline{u}^{\text{i}} - \mathcal{S}_f \overline{v}^{\text{i}} \right). \qquad (6.36)$$

In obtaining the third line, the meridional derivative of $\epsilon_{\text{ek,bot}}$ is ignored for simplicity. (Because ν_{e} is a sufficiently uncertain closure parameter, its horizontal derivative should be considered to be even more uncertain, and when in doubt leave it out; cf. Occam's Razor.) In the final line the horizontal divergence of $\mathbf{u}^{\text{i}}$ is neglected relative to its vorticity, ζ^{i}, based on an assumption that $Ro \ll 1$ for the interior flow, consistent with its geostrophic balance in (6.17) (Section 2.4.2).

6.1.5 Oceanic surface Ekman layer

The preceding analysis (6.6)–(6.36) has been for a fluid above a solid boundary, appropriate to the bottom of the ocean and atmosphere. Even for winds over the fluid ocean, the surface currents are usually so much slower than the winds that the no-slip condition is a good approximation for the atmospheric dynamics. However, as mentioned near the end of Section 6.1.1, an analogous analysis can be made for the boundary on top of the fluid, appropriate to the top of the ocean forced by a surface wind stress with $\rho_o \overline{\mathbf{u}' w'} = -\boldsymbol{\tau}_{\text{s}}$ at the boundary. In particular, the surface stress (6.32) that results from the atmospheric boundary layer beneath the atmospheric interior wind, $\overline{\mathbf{u}}^{\text{i}}$, is used to force the oceanic planetary boundary layer. Now, however, $\boldsymbol{\tau}_{\text{s}}$ must be interpreted as the drag of the boundary on the underlying fluid, rather than vice versa in (6.13) and (6.14), due to the reversal in the sign of z relative to the boundary location. Since the oceanic derivation parallels Section 6.1.4, the primary relations for an oceanic laminar surface Ekman layer can be briefly summarized as the following:

$$\nu_{\text{e}} \frac{\partial^2 U}{\partial z^2} = i f U$$

$$\nu_{\text{e}} U_z(0) = \frac{1}{\rho_0} (\tau_{\text{s}}^x + i \tau_{\text{s}}^y)$$

$$U(-\infty) = 0, \qquad (6.37)$$

and

$$U(z) = (1 - i\mathcal{S}_f) \frac{\tau_s^x + i\tau_s^y}{\rho_0\sqrt{2f\nu_e}} e^{\lambda(1+i\mathcal{S}_f)z}$$

$$\overline{u}(z) = \frac{1}{\rho_0\sqrt{2|f|\nu_e}} e^{\lambda z}((\tau_s^x + \mathcal{S}_f\tau_s^y)\cos[\lambda z]$$

$$+ (\tau_s^x - \mathcal{S}_f\tau_s^y)\sin[\lambda z])$$

$$\overline{v}(z) = \frac{1}{\rho_0\sqrt{2|f|\nu_e}} e^{\lambda z}((-\mathcal{S}_f\tau_s^x + \tau_s^y)\sin[\lambda z]$$

$$+ (\mathcal{S}_f\tau_s^x + \tau_s^y)\cos[\lambda z]), \tag{6.38}$$

and

$$T^x = \int_{-\infty}^{0} \overline{u}^b dz = \frac{\tau_s^y}{\rho_0 f}$$

$$T^y = \int_{-\infty}^{0} \overline{v}^b dz = -\frac{\tau_s^x}{\rho_0 f}, \tag{6.39}$$

and

$$\overline{w}^b(-\infty) = w_{ek,top} = \hat{\mathbf{z}} \cdot \nabla \times \frac{\tau_s}{\rho_0 f}. \tag{6.40}$$

The oceanic boundary-layer transport is rotated 90° to the right (left) of the surface wind stress exerted at the boundary on the ocean for $f > 0$ ($f < 0$) in the northern (southern) hemisphere. The stress is parallel to the wind direction just above the air–sea interface. The Ekman spiral in the ocean starts with a surface velocity 45° to the right (left) of the surface stress, and it decreases in magnitude and rotates clockwise (counter-clockwise) in direction with increasing depth until it vanishes to zero in the deep interior. The oceanic surface current is therefore in the same direction as the atmospheric interior wind in these solutions using the eddy viscosity closure (6.23). These relations are illustrated in Fig. 6.4.

A spatially broader view of the relations among the local wind, surface stress, current, boundary-layer transports, and Ekman pumping is sketched in Fig. 6.5 for the situation of a tropospheric cyclone above the ocean (neglecting the β term in (6.36) for simplicity). Note that the atmospheric Ekman pumping acts to flux fluid from the planetary boundary layer upward into the central interior of the cyclone, supported by an inward radial Ekman transport. This has the effect of increasing its hydrostatic pressure there and, since a cyclone has a low central geostrophic pressure (Section 3.1.4), this provides a tendency for weakening the cyclone and spinning down its circulation (a further analysis is made in Section 6.1.6). Conversely, the upward Ekman pumping in the oceanic planetary boundary layer,

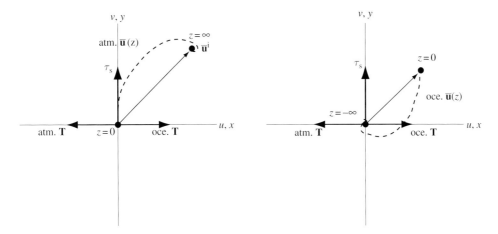

Fig. 6.4. Oceanic and atmospheric surface Ekman layers in the northern hemisphere. The atmospheric layer (left) has a left-turning spiral in $\bar{u}^a(z)$ approaching the boundary and a boundary-layer transport anomaly, $\mathbf{T}_{atm.}$, directed $135°$ to the left of the interior wind. The oceanic layer (right) has a surface current, $\bar{u}^o(z)$, directed $45°$ to the right of the surface stress, τ_s (which itself is $45°$ to the left of the interior wind), a right-turning spiral in $\bar{u}^o(z)$ going away from the boundary and a transport, $\mathbf{T}_{oce.}$, directed $90°$ to the right of the surface stress. Note the equal and opposite transports in the atmospheric and oceanic layers.

supported by its outward radial Ekman transport, acts through vortex stretching (analogous to the shallow-water example in Section 4.1.1) to reduce its central pressure by spinning up an oceanic geostrophic cyclonic circulation underneath the tropospheric one. This is an example of dynamical coupling between the atmosphere and ocean through the turbulent surface drag stress.

Since the oceanic surface Ekman layer problem (6.37) is independent of the total surface velocity, because the contribution of the current to determining the surface stress is neglected, its Ekman current can be superimposed on any interior geostrophic current profile. (This property is further exploited in Section 6.2 for an oceanic wind gyre.) Alternatively expressed, the shear instability that drives the oceanic planetary boundary-layer turbulence is usually due to the mean boundary-layer shear, not the interior geostrophic shear, even when the latter is baroclinic in a stratified ocean.

Coastal upwelling and downwelling

Although the climatological winds over the ocean are primarily zonal (Section 6.2), there are some locations where they are more meridional and parallel to the continental coastline. This happens for the dominant extra-tropical, marine standing eddies in the atmosphere, anticyclonic *sub-tropical highs* and cyclonic *sub-polar lows* (referring to their surface pressure extrema). On the

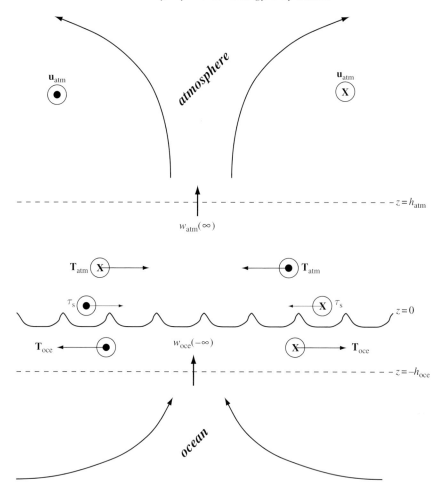

Fig. 6.5. Sketch of the boundary-layer depths (h_{atm} and h_{oce}), horizontal transports ($\mathbf{T}_{atm}$ and $\mathbf{T}_{oce}$), Ekman pumping (w_{atm} and w_{oce}), and interior ageostrophic flows (curved arrows) for an atmospheric cyclone, $\mathbf{u}_{atm}$, over the ocean (northern hemisphere). The interior and boundary-layer velocity component perpendicular to the plotted cross-sectional plane are indicated by a dot or cross within a circle (flow out of or into the plane, respectively).

eastern side of these atmospheric eddies and adjacent to the eastern boundary of the underlying oceanic basins, the surface winds are mostly equatorward in the sub-tropics and poleward in the sub-polar zone. The associated surface Ekman transports (6.39) are off-shore and on-shore, respectively. Since the along-shore scale of the wind is quite large ($\sim$1000s km) and the normal component of current must vanish at the shoreline, incompressible mass balance requires that the water comes up from below or goes downward near the coast, within $\sim$10s km. This circulation pattern is called *coastal upwelling* or *downwelling*. It is a prominent feature in the oceanic general circulation (Figs. 1.1 and 1.2). It also has the important biogeochemical consequence of fueling high plankton productivity: upwelling brings chemical nutrients (e.g., nitrate) to the surface layer where there is abundant sunlight; examples are the sub-tropical Benguela Current off South Africa and the California Current off North America. Analogous behavior can occur adjacent to other oceanic basin boundaries, but many typically have winds less parallel to the coastline, hence weaker upwelling or downwelling.

6.1.6 Vortex spin down

The bottom Ekman pumping relation in (6.36) implies a *spin down* (i.e., decay in strength) for the overlying interior flow. Continuing with the assumptions that the interior has uniform density and its flow is approximately geostrophic and hydrostatic, then the Taylor–Proudman theorem (6.15) implies that the horizontal velocity and vertical vorticity are independent of height, while the interior vertical velocity is a linear function of depth. Assume the interior layer spans $0 < h \leq z \leq H$, where $w = \overline{w}^{\mathrm{i}} + \overline{w}^{\mathrm{b}}$ has attained its Ekman pumping value (6.36) by $z = h$ and vanishes at the top height, $z = H$. $\overline{w}^{\mathrm{i}}$ from (6.18) is small compared to $\overline{w}^{\mathrm{b}} = w_{\mathrm{ek}}$ at $z = h$ since $h \ll H$ and $\overline{w}^{\mathrm{i}}(H) = -w_{\mathrm{ek}}$.

An axisymmetric vortex on the f-plane (Section 3.1.4) has no evolutionary tendency associated with its azimuthal advective nonlinearity, but the vertical velocity does cause the vortex to change with time according to the barotropic vorticity equation for the interior layer:

$$\frac{\partial \overline{\zeta}^{\mathrm{i}}}{\partial t} = -(f + \overline{\zeta}^{\mathrm{i}})\nabla_{\mathrm{h}} \cdot \overline{\mathbf{u}}_{\mathrm{h}}^{\mathrm{i}} \approx -f\nabla_{\mathrm{h}} \cdot \overline{\mathbf{u}}_{\mathrm{h}}^{\mathrm{i}}$$

$$= f\frac{\partial \overline{w}}{\partial z}$$

$$= f\left(\frac{\overline{w}(H) - \overline{w}(h)}{H - h}\right) = -\left(\frac{f}{H - h}\right)\overline{w}^{\mathrm{b}}$$

$$= -\left(\frac{\epsilon_{\mathrm{ek,bot}}}{H - h}\right)\overline{\zeta}^{\mathrm{i}}, \tag{6.41}$$

where $\overline{\zeta}^{\mathrm{i}}$ is neglected relative to f in the first line by assuming small Ro. This equation is readily integrated in time to give

$$\overline{\zeta}^{\mathrm{i}}(r, t) = \overline{\zeta}^{\mathrm{i}}(r, 0)e^{-t/t_{\mathrm{d}}}. \tag{6.42}$$

This result shows that the vortex preserves its radial shape, while decaying in strength, with a spin-down time defined by

$$t_{\mathrm{d}} = \frac{H - h}{\epsilon_{\mathrm{ek,bot}}} = \sqrt{\frac{2(H - h)^2}{|f|\nu_{\mathrm{e}}}} \approx \frac{1}{f\sqrt{E}}. \tag{6.43}$$

In the last relation in (6.43), a non-dimensional *Ekman number* is defined by

$$E = \frac{2\nu_e}{f_0 H^2}. \tag{6.44}$$

E is implicitly assumed to be small since

$$h_{ek} = \sqrt{\frac{2\nu_e}{f_0}} = \sqrt{E} H \ll H \iff E \ll 1$$

is a necessary condition for this kind of vertical boundary-layer analysis to be valid (Section 6.1.1). Therefore, the vortex spin down time is much longer than the Ekman-layer set-up time, $\sim 1/f$. Consequently, the Ekman layer evolves in a quasi-steady balance, keeping up with the interior flow as the vortex decays in strength.

For strong vortices such as a hurricane, this type of analysis for a quasi-steady Ekman layer and axisymmetric vortex evolution can be generalized with gradient-wind balance (rather than geostrophic, as above). The results are that the vortex spins down with a changing radial shape (rather than an invariant one); a decay time, t_d, that additionally depends upon the strength of the vortex; and an algebraic (rather than exponential) functional form for the temporal decay law (Eliassen and Lystad, 1977). Nevertheless, the essential phenomenon of vortex decay is captured in the linear model (6.41).

6.1.7 Turbulent Ekman layer

The preceding Ekman layer solutions are all based on the boundary-layer approximation and eddy viscosity closure, whose accuracies need to be assessed. The most constructive way to make this assessment is by *direct numerical simulation* (DNS) of the governing equations (6.1), with uniform $f = f_0$; a Newtonian viscous diffusion (6.10) with large Re; an interior barotropic, geostrophic velocity, $\mathbf{u}^i$; a no-slip bottom boundary condition at $z = 0$; an upper boundary located much higher than $z = h$; a horizontal boundary condition of periodicity over a spatial scale, L, again much larger than h; and a long enough integration time to achieve a statistical equilibrium state. This simulation provides a uniform-density, homogeneous, stationary truth standard for assessing the Ekman boundary-layer and closure approximations.

An alternative to DNS and mean-field closure models (e.g., Section 6.1.3) is *large-eddy simulation* (LES). LES is an intermediate level of dynamical approximation in which the fluid equations are solved with non-conservative eddy-flux divergences representing transport by turbulent motions on scales smaller than those resolved with the computational discretization (rather than by all the turbulence as in a mean-field model). These *sub-grid-scale fluxes* must be specified by a closure theory expressed as a parameterization (Chapter 1), whether as simple as eddy diffusion or more elaborate. The turbulent flow simulations using eddy viscosities in Sections 5.3 and 6.2.4 can therefore be considered as examples of LES, as can general circulation models. LES is also commonly applied to planetary boundary layers, often with a somewhat elaborate parameterization.

A numerical simulation requires a discretization of the governing equations onto a spatial grid. The grid dimension, N, is then chosen to be as large as possible on the computer available so that Re can be as large as possible to mimic geophysical boundary layers. The grid spacing, e.g., $\Delta x = L/N$, is determined by the requirement that the viscous term – with the highest order of spatial differentiation, hence the finest scales of spatial variability (Section 3.7) – be well resolved. This means that the solution is spatially smooth between neighboring grid points, and in practice this occurs only if a grid-scale Reynolds number is not too large,

$$Re_g = \frac{\Delta V \Delta x}{\nu} = \mathcal{O}(1),$$

where Δ denotes differences on the grid scale. For a planetary boundary-layer flow, this is equivalent to the requirement that the near-surface, viscous sub-layer be well resolved by the grid. The value of the macro-scale $Re = VL/\nu$ is then chosen to be as large as possible, by making $(V/\Delta V) \cdot (L/\Delta x)$ as large as possible. Present computers allow calculations with $Re = \mathcal{O}(10^3)$ for isotropic, 3D turbulence. Although this is nowhere near the true geophysical values for the planetary boundary layer, it is large enough to lie within what is believed to be the regime of fully developed turbulence. With the hypothesis that Re dependences for fully developed turbulence are merely quantitative rather than qualitative and associated more with changes on smaller scales than with the *energy-containing scale*, $\sim h$, that controls the Reynolds stress and velocity variance, then the results of these feasible numerical simulations are relevant to the natural planetary boundary layers.

The $\overline{\mathbf{u}}(z)$ profile calculated from the solution of such a direct numerical simulation with $f > 0$ is shown in Fig. 6.6. It has a shape qualitatively similar to the laminar Ekman layer profile (Section 6.1.3). The surface current is rotated to the left of the interior current, though by less than the 45° of the laminar profile (Fig. 6.7), and the currents spiral with height, though less strongly so than in the

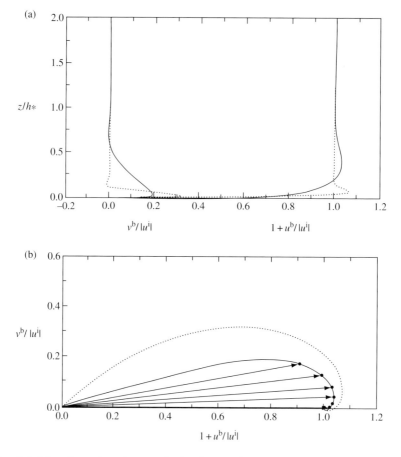

Fig. 6.6. Mean boundary-layer velocity for a turbulent Ekman layer at $Re = 10^3$. Axes are aligned with $\mathbf{u}^i$. (a) profiles with height; (b) hodograph. The solid lines are for the numerical simulation, and the dashed lines are for a comparable laminar solution with a constant eddy viscosity, ν_e (Coleman, 1999).

laminar Ekman layer. Of course, the transport, $\mathbf{T}$, must still satisfy (6.21). The vertical decay scale, h_*, for $\bar{\mathbf{u}}(z)$ is approximately

$$h_* = 0.25 \, \frac{u_*}{f}, \tag{6.45}$$

where

$$u_* = \sqrt{\frac{|\tau_s|}{\rho_0}} \tag{6.46}$$

is the *friction velocity* based on the surface stress. In a gross way this can be compared to the laminar decay scale, $\lambda^{-1} = \sqrt{2\nu_e/f}$, from (6.29). The two length

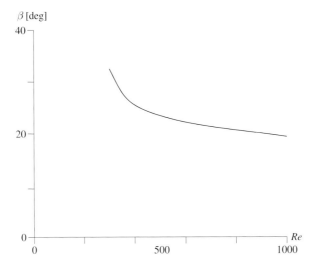

Fig. 6.7. Sketch of clockwise rotated angle, β, of the surface velocity relative to $\mathbf{u}^i$ as a function of Re within the regime of fully developed turbulence, based on 3D computational solutions. For comparison, the laminar Ekman layer value is $\beta = 45°$. (Adapted from Coleman, 1999.)

scales are equivalent for an eddy viscosity value of

$$\nu_e = 0.03 \, \frac{u_*^2}{f} = 0.13 \; u_* h_*. \tag{6.47}$$

The second relation is consistent with the widespread experience that eddy viscosity magnitudes diagnosed from the negative of the ratio of eddy flux and the mean gradient (6.23) are typically a small fraction of the product of an eddy speed, V', and an eddy length scale, L'. An eddy viscosity relation of this form, with

$$\nu_e \sim V' L', \tag{6.48}$$

is called a *mixing-length* estimate. Only after measurements or turbulent simulations have been made are u_* and h_* (or V' and L') known, so that an equivalent eddy viscosity (6.47) can be diagnosed.

The turbulent and viscous stress profiles (Fig. 6.8) show a rotation and decay with height on the same boundary-layer scale, h_*. The viscous stress is negligible compared to the Reynolds stress except very near the surface. Near the surface within the viscous sub-layer, the Reynolds stress decays to zero, as it must, because of the no-slip boundary condition, and the viscous stress balances the Coriolis force in equilibrium, allowing the interior mean velocity profile to smoothly continue to its boundary value. By evaluating (6.23) locally at any height, the ratio of turbulent stress and mean shear is equal to the diagnostic eddy viscosity, $\nu_e(z)$. Its characteristic profile is sketched in Fig. 6.9. It has a convex shape. Its

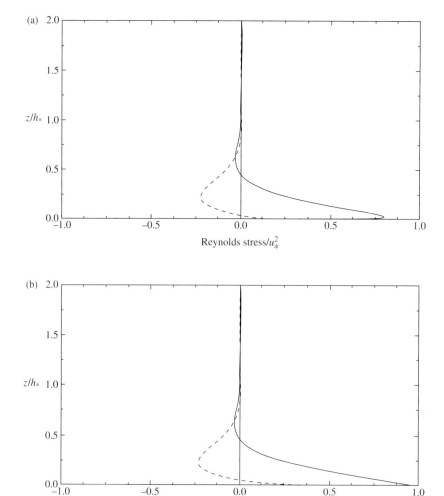

Fig. 6.8. Momentum flux (or stress) profiles for a turbulent Ekman layer at $Re = 10^3$. Axes are aligned with $\mathbf{u}^i$. (a) $-\overline{\mathbf{u}'w'}(z)$; (b) Reynolds plus viscous stress. The solid line is for the streamwise component, and the dashed line is for the cross-stream component. Note that the Reynolds stress vanishes very near the surface within the viscous sub-layer, while the total stress is finite there (adapted from Coleman, 1999).

peak value is in the middle of the planetary boundary layer and is several times larger than the gross estimate (6.47). It decreases toward both the interior and the solid surface. It is positive everywhere, implying a down-gradient momentum flux by the turbulence. Thus, the diagnosed eddy viscosity is certainly not the constant value assumed in the laminar Ekman layer (Section 6.1.4), but neither does it wildly deviate from it.

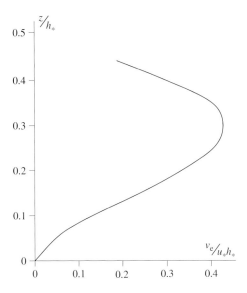

Fig. 6.9. Sketch of eddy viscosity profile, $\nu_e(z)$, for a turbulent Ekman layer. Note the convex shape with smaller ν_e near the boundary and approaching the interior.

The diagnosed $\nu_e(z)$ indicates that the largest discrepancies between laminar and turbulent Ekman layers occur near the solid-boundary and interior edges. The boundary edge is particularly different. In addition to the thin viscous sub-layer, where all velocities smoothly go to zero as $z \to 0$, there is an intermediate turbulent layer called the *log layer* or *similarity layer*. Here the important turbulent length scale is not the boundary-layer thickness, h_*, but the distance from the boundary, z. In this layer the mean velocity profile has a large shear with a profile shape governed by the boundary stress (u_*) and the near-boundary turbulent eddy size (z) in the following way:

$$\frac{\partial \overline{\mathbf{u}}}{\partial z} = K \frac{u_*}{z} \, \hat{\mathbf{s}}$$

$$\implies \overline{\mathbf{u}}(z) = K \, u_* \, \ln\left[\frac{z}{z_o}\right] \hat{\mathbf{s}}. \tag{6.49}$$

This is derived by *dimensional analysis*, a variant of the scaling analyses frequently used above, as the only dimensionally consistent combination of only u_* and z, with the implicit assumption that Re is irrelevant for the log layer (as $Re \to \infty$). In (6.49) $K \approx 0.4$ is the empirically determined *von Karman constant*; z_o is an integration constant called the *roughness length* that characterizes the irregularity of the underlying solid surface; and $\hat{\mathbf{s}}$ is a unit vector in the direction of the surface stress. Measurements show that K does not greatly vary from one natural situation to another, but z_o does. The logarithmic shape for $\bar{\mathbf{u}}(z)$ in (6.49) is the basis for the name of this intermediate layer. The log layer quantities have no dependence on f, hence they are not a part of the laminar Ekman layer paradigm (Sections 6.1.3–6.1.5), which is thus more germane to the rest of the boundary layer above the log layer.

In a geophysical planetary boundary-layer context, the log layer is also called the *surface layer*, and it occupies only a small fraction of the boundary-layer height, h (e.g., typically 10–15%). (This is quite different from non-rotating shear layers where the profile (6.49) extends throughout most of the turbulent boundary layer.) Figure 6.10 is a sketch of the near-surface mean velocity profile, and it shows the three different vertical layers in the turbulent shear planetary boundary layer: viscous sub-layer, surface layer, and Ekman boundary layer. In natural planetary boundary layers with stratification, the surface similarity layer profile (6.49) also occurs but in a somewhat modified form (often called *Monin–Obukhov similarity*). Over very rough lower boundaries (e.g., in the atmosphere above a forest canopy or a field of surface gravity waves), the similarity layer shifts to somewhat greater heights, well above the viscous sub-layer, and the value of z_o increases substantially; furthermore, the surface stress, $\boldsymbol{\tau}_{\mathrm{s}}$, is dominated by form stress due to pressure forces on the rough boundary elements (Section 5.3.3) rather than viscous stress.

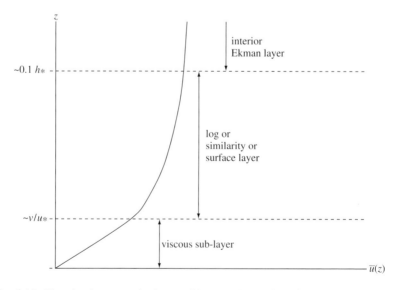

Fig. 6.10. Sketch of mean velocity profile near the surface for a turbulent Ekman layer. Note the viscous sub-layer and the logarithmic (also known as surface or similarity) layer that occur closer to the boundary than the Ekman spiral in the interior region of the boundary layer.

Under the presumption that the Reynolds stress profile approaches the boundary smoothly on the vertical scale of the Ekman layer (Fig. 6.8), a diagnostic eddy viscosity profile (6.23) in the log layer must have the form of

$$\nu_{\mathrm{e}}(z) = \frac{u_* z}{K}. \tag{6.50}$$

This is also a mixing-length relationship (6.48) constructed from a dimensional analysis with $V' \sim u_*$ and $L' \sim z$. $\nu_{\mathrm{e}}(z)$ vanishes as $z \to 0$, consistent with the shape sketched in Fig. 6.9. The value of $\nu_{\mathrm{e}}(z)$ in the log layer (6.50) is smaller than its gross value in the Ekman layer (6.47) as long as z/h_* is less than about 0.05, i.e., within the surface layer.

The turbulent Ekman layer problem has been posed here in a highly idealized way. Usually in natural planetary boundary layers there are important additional influences from density stratification and surface buoyancy fluxes; the horizontal component of the Coriolis vector (Section 2.4.2); and the variable topography of the bounding surface, including the moving boundary for air flow over surface gravity waves and wave-averaged Stokes-drift effects (Section 3.5) in the oceanic boundary layer.

6.2 Oceanic wind gyre and western boundary layer

Consider the problem of a mid-latitude *oceanic wind gyre* driven by surface wind stress over a zonally bounded domain. This is the prevailing form of the oceanic general circulation in mid-latitude regions, excluding the ACC south of 50° S. A wind gyre is a horizontal recirculation cell spanning an entire basin, i.e., with the largest scale of $5-10 \times 10^3$ km. The sense of the circulation is anticyclonic in the sub-tropical zones (i.e., the latitude band of 20°–45°) and cyclonic in the sub-polar zones (45°–65°); Fig. 6.11. This gyre structure is a forced response to the general pattern of the mean surface zonal winds (Fig. 5.1): tropical easterly trade winds, extra-tropical westerlies, and weak or easterly polar winds.

This problem involves the results of both the preceding Ekman layer analysis and a *western boundary current* that is a lateral, rather than vertical, boundary layer within a wind gyre with a much smaller lateral scale, $<10^2$ km, than the gyre itself. This problem was first posed and solved by Stommel (1948) in a highly simplified form (Sections 6.2.1 and 6.2.2). It has been extensively studied since then – almost as often as the zonal baroclinic jet problem in Section 5.3 – because it is such a central phenomenon in oceanic circulation and because it has an inherently turbulent, eddy–mean interaction in statistical equilibrium (Section 6.2.4). The wind gyre is yet another perennially challenging GFD problem.

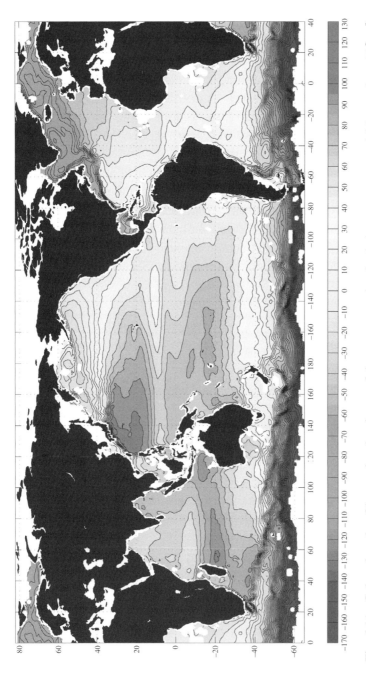

Fig. 6.11. Color plate 6. Observational estimate of time-mean sea level relative to a geopotential iso-surface, $\overline{\eta}$ [cm]. The estimate is based on near-surface drifting buoy trajectories, satellite altimetric heights, and climatological winds. $g\overline{\eta}/f$ can be interpreted approximately as the surface geostrophic streamfunction. Note the sub-tropical and sub-polar wind gyres with sea-level extrema adjacent to the continental boundaries on western sides of the major basins and the large sea-level gradient across the Antarctic Circumpolar Current (Niiler *et al.*, 2003).

6.2.1 Posing the gyre problem

The idealized wind-gyre problem is posed for a uniform density ocean in a rectangular domain with a rigid lid (Section 2.2.3) and a steady zonal wind stress at the top,

$$\boldsymbol{\tau}_s = \tau_s^x(y)\,\hat{\mathbf{x}} \tag{6.51}$$

(Fig. 6.12). Make the β-plane approximation (Section 2.4) and assume the gyre is in the northern hemisphere (i.e., $f > 0$). Also assume that there are Ekman boundary layers both near the bottom at $z = 0$, where $\mathbf{u} = 0$ as in Sections 6.1.2–6.1.4, and near the top at $z = H$ with an imposed stress (6.51) as in Section 6.1.5. Thus, the ocean is split into three layers (Fig. 6.13). These are the interior layer between the two boundary layers, and the latter are much thinner than the ocean as a whole. Based on an eddy-viscosity closure for the vertical boundary layers (Section 6.1.3) and the assumption that the Ekman number, E in (6.44), is small, then an analysis for the interior flow can be made to be similar to the problem of vortex spin down (Section 6.1.6).

Within the interior layer, the 3D momentum balance is approximately geostrophic and hydrostatic. A scale estimate with $V = 0.1\,\mathrm{m\ s^{-1}}$, $L = 5 \times 10^6\,\mathrm{m}$, $H = 5\,\mathrm{km}$, and $f = 10^{-4}\,\mathrm{s^{-1}}$ implies a Rossby number of $Ro = 0.5 \times 10^{-4} \ll 1$ and an aspect ratio of $H/L = 10^{-3} \ll 1$. So these approximations are well founded. Because of the Taylor–Proudman theorem (6.15), the horizontal velocity and horizontal pressure gradient must be independent of depth (i.e., barotropic) within the interior layer, and because of 3D continuity, the vertical velocity is at most a linear function of z in the interior. The Ekman layers are not constrained by the

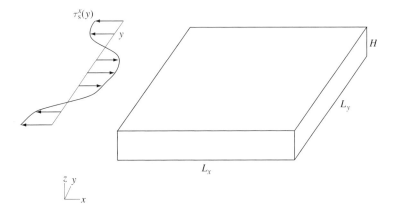

Fig. 6.12. Oceanic gyre domain shape and surface zonal wind stress, $\tau_s^x(y)$. The domain is rectangular with a flat bottom (i.e., $L_x \times L_y \times H$). The density is uniform.

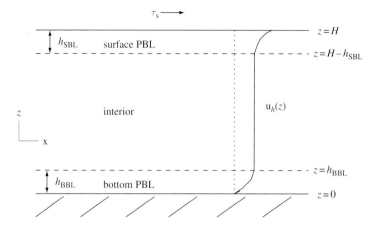

Fig. 6.13. Vertical layers for a uniform-density oceanic wind gyre. The interior, hydrostatic, geostrophic, horizontal velocity is independent of depth, and there are surface and bottom Ekman layers to accommodate the surface stress and no-slip boundary conditions, respectively.

Taylor–Proudman theorem since their large turbulent Reynolds stress makes the momentum balance (6.19) and (6.20) ageostrophic.

The relevant Ekman layer properties are the top and bottom horizontal transports,

$$\mathbf{T}_{\text{ek, top}} = -\hat{\mathbf{z}} \times \frac{\tau_s}{\rho_0 f}$$

$$\mathbf{T}_{\text{ek, bot}} = \frac{\epsilon_{\text{ek, bot}}}{f} \left(-u_{\text{bot}}^{\text{i}} - v_{\text{bot}}^{\text{i}}, \ u_{\text{bot}}^{\text{i}} - v_{\text{bot}}^{\text{i}} \right) , \tag{6.52}$$

and Ekman pumping,

$$w_{\text{ek, top}} = \hat{\mathbf{z}} \cdot \boldsymbol{\nabla} \times \left[\frac{\tau_s^x \hat{\mathbf{x}}}{\rho_0 f} \right] = -\frac{1}{\rho_0} \frac{\partial}{\partial y} \left[\frac{\tau_s^x}{f} \right]$$

$$w_{\text{ek, bot}} = \frac{\epsilon_{\text{ek, bot}}}{f} \zeta_{\text{bot}}^{\text{i}} + \frac{\beta \epsilon_{\text{ek, bot}}}{f^2} \left(u_{\text{bot}}^{\text{i}} - v_{\text{bot}}^{\text{i}} \right) \tag{6.53}$$

(Sections 6.1.4 and 6.1.5). The subscripts "ek, top" and "ek, bot" denote the surface and bottom Ekman boundary layers, respectively, and the superscript "i" denotes the interior value outside the boundary layer.

In the interior the flow is barotropic. Therefore, the depth-averaged vorticity for the interior region is simply the vorticity itself,

$$\frac{1}{H_{\text{I}}} \int_{h_{\text{ek, bot}}}^{H - h_{\text{ek, top}}} \zeta \, dz = \zeta^{\text{i}}(x, y, t) \quad \text{and}$$

$$\zeta(z = h_{\text{ek, bot}}) = \zeta^{\text{i}}. \tag{6.54}$$

ζ^{i} is the interior relative vorticity, and

$$H_{\mathrm{I}} = H - h_{\mathrm{ek,\ top}} - h_{\mathrm{ek,\ bot}} \approx H$$

is the thickness of the interior region. The depth-averaged vorticity equation (cf., (3.24)) can be written in the interior as

$$\frac{\partial \zeta^{\mathrm{i}}}{\partial t} + \mathbf{u}_{\mathrm{h}}^{\mathrm{i}} \cdot \nabla (f(y) + \zeta^{\mathrm{i}}) = -(f + \zeta^{\mathrm{i}}) \nabla \cdot \mathbf{u}_{\mathrm{h}}^{\mathrm{i}} + \mathcal{F}^{\mathrm{i}}. \tag{6.55}$$

Use the Ekman pumping relations (6.53) and the continuity equation to evaluate the planetary vortex stretching (i.e., the first right-hand side term in (6.55):

$$-f \nabla \cdot \mathbf{u}_{\mathrm{h}}^{\mathrm{i}} = f \frac{\partial w^{\mathrm{i}}}{\partial z}$$

$$= \frac{f}{H} \left(w_{\mathrm{ek,\ top}} - w_{\mathrm{ek,\ bot}} \right)$$

$$= -\frac{1}{\rho_0 H} \frac{\partial \tau_{\mathrm{s}}^x}{\partial y} + \frac{\beta \tau_{\mathrm{s}}^x}{f \rho_0 H}$$

$$- \frac{\epsilon_{\mathrm{ek,\ bot}}}{H} \left(\zeta^{\mathrm{i}} - \frac{\beta}{f} (u^{\mathrm{i}} - v^{\mathrm{i}}) \right). \tag{6.56}$$

Equation (6.55) is further simplified here by the additional assumptions that the flow is steady in time ($\partial_t \zeta^{\mathrm{i}} = 0$); that the interior non-conservative term, $\mathcal{F}^{\mathrm{i}}$, is negligible; and the flow is weak enough that the nonlinear terms are also negligible. The result is

$$\beta H v^{\mathrm{i}} = -\frac{1}{\rho_0} \frac{\partial \tau_{\mathrm{s}}^x}{\partial y} + \frac{\beta \tau_{\mathrm{s}}^x}{f \rho_0} - \epsilon_{\mathrm{ek,\ bot}} \zeta^{\mathrm{i}} - \frac{\beta \epsilon_{\mathrm{ek,\ bot}}}{f} (u^{\mathrm{i}} - v^{\mathrm{i}}). \tag{6.57}$$

This is a formula for the interior meridional transport, $H v^{\mathrm{i}}$. To obtain the expressions for the total meridional transport, $T^y = H V$, namely,

$$H V = H v^{\mathrm{i}} + T_{\mathrm{ek,\ top}}^y + T_{\mathrm{ek,\ bot}}^y,$$

multiply the T_{ek}^y expressions in (6.52) by β and add them to (6.57):

$$\beta \left(H v^{\mathrm{i}} + T_{\mathrm{ek,\ top}}^y + T_{\mathrm{ek,\ bot}}^y \right) = -\frac{1}{\rho_0} \frac{\partial \tau_{\mathrm{s}}^x}{\partial y} + \frac{\beta \tau_{\mathrm{s}}^x}{f \rho_0} - \epsilon_{\mathrm{ek,\ bot}} \zeta^{\mathrm{i}}$$

$$- \frac{\beta \epsilon_{\mathrm{ek,\ bot}}}{f} (u^{\mathrm{i}} - v^{\mathrm{i}}) - \beta \frac{\tau_{\mathrm{s}}^x}{\rho_0 f} + \beta \frac{\epsilon_{\mathrm{ek,\ bot}}}{f} (u^{\mathrm{i}} - v^{\mathrm{i}})$$

$$\implies \beta H V = -\frac{1}{\rho_0} \frac{\partial \tau_{\mathrm{s}}^x}{\partial y} - \epsilon_{\mathrm{ek,\ bot}} \zeta^{\mathrm{i}}. \tag{6.58}$$

As the final step in the derivation of the barotropic wind-gyre equation, vertically integrate the continuity equation with the kinematic boundary condition for flat top and bottom surfaces, i.e., $w = 0$. The result is that the depth-averaged horizontal velocity (U, V) is horizontally non-divergent. Hence a *transport stream-function*, $\Psi(x, y)\,[\mathrm{m^3\,s^{-1}}]$, can be defined by

$$T^x = HU = -\frac{\partial \Psi}{\partial y}, \quad T^y = HV = \frac{\partial \Psi}{\partial x}. \tag{6.59}$$

Ψ differs from the usual streamfunction, ψ, by an added depth integration, so $\Psi = H\psi$.

Since the bottom Ekman-layer velocity has about the same magnitude as the interior velocity (to satisfy the no-slip condition at the bottom), the bottom transport will be small compared to the interior transport by the ratio, $h_{\mathrm{ek,\ bot}}/H = E^{1/2} \ll 1$. On the other hand, the surface Ekman and interior transports are comparable, with the surface Ekman velocity much larger than the interior velocity. The curl of the total transport (6.59), neglecting the $\mathcal{O}(E^{1/2})$ bottom Ekman-layer contribution, gives the relation,

$$\nabla^2 \Psi \approx H\zeta^{\mathrm{i}} + \hat{\mathbf{z}} \cdot \nabla \times \frac{\tau_{\mathrm{s}}^x \hat{\mathbf{x}}}{\rho_0 f}.$$

Using this relation for ζ^{i} and (6.59) for V and substituting into (6.58) yields the steady, linear, barotropic potential-vorticity equation,

$$\frac{\epsilon_{\mathrm{ek,\ bot}}}{H}\nabla^2 \Psi + \beta\frac{\partial \Psi}{\partial x} = -\frac{1}{\rho_0}\frac{\partial \tau_{\mathrm{s}}^x}{\partial y} + \frac{\epsilon_{\mathrm{ek,\ bot}}}{H}\,\hat{\mathbf{z}} \cdot \nabla \times \frac{\tau_{\mathrm{s}}^x \hat{\mathbf{x}}}{\rho_0 f}$$

$$\approx -\frac{1}{\rho_0}\frac{\partial \tau_{\mathrm{s}}^x}{\partial y}. \tag{6.60}$$

The second right-hand side term in the first line is $\mathcal{O}(\epsilon_{\mathrm{ek,\ bot}}/fH) = \mathcal{O}(E^{1/2})$ relative to the first one by (6.33)–(6.44), hence negligibly small. Equation (6.60) will be solved in Section 6.2.2.

Discussion

As an alternative path to the same result, the gyre equations (6.58)–(6.60) could be derived directly and more concisely from a vertical integral of the steady, linear, conservative, mean-field vorticity equation,

$$\beta v = f\frac{\partial w}{\partial z} - \hat{\mathbf{z}} \cdot \nabla \times \frac{\partial}{\partial z}\overline{w'\mathbf{u}'}$$

using the appropriate kinematic and stress boundary conditions. But the preceding derivation is preferable because it emphasizes the role of the Ekman boundary layers as depicted in Fig. 6.13.

It is noteworthy that the resulting vorticity equation (6.58) has apparent body forces that are equivalent to those contained in the non-conservative forces, $\mathbf{F}_n$, for an N-layer quasigeostrophic model (Section 5.3). To see this, take the curl of (5.80), perform a discrete vertical integration (to match the depth-integrated, single-layer situation here), and neglect the eddy viscosity contributions. The result is the equivalence relation,

$$\sum_{n=1}^{N} H_n \hat{\mathbf{z}} \cdot \nabla \times \mathbf{F}_n = -\frac{1}{\rho_o}\frac{\partial \tau_s^x}{\partial y} - \epsilon_{\text{bot}}\,\zeta_{\text{bot}}.$$

This expression is equal to the right-hand side of (6.58) for the particular choice of the bottom-drag coefficient,

$$\epsilon_{\text{bot}} = H f_0 \sqrt{\frac{E}{2}} = \sqrt{\frac{\nu_e f_0}{2}},$$

consistent with (6.33) and (6.44).

Therefore, there is an equivalence between two different conceptions of a layered quasigeostrophic model with vertical boundary stresses.

• Explicitly resolve the Ekman layers between the vertical boundaries and the adjacent interior quasigeostrophic layers, $n = 1$ and N (as done here), with the Ekman transport, $\mathbf{T}_{\text{ek}}$, added to the interior ageostrophic horizontal transport, $H_n \mathbf{u}_{\text{a},n}$, and the Ekman pumping, w_{ek}, contributing to the vortex stretching in the vorticity and potential-vorticity equations.

• Implicitly embed the Ekman layers within the layers $n = 1$ and N through an equivalent body force in $\mathbf{F}_n$ (as in Section 5.3.1); the consequences are that the Ekman transport is a depth-weighted fraction, $\mathbf{T}_{\text{ek}}/H_n$, of the layer ageostrophic flow, $H_n \mathbf{u}_{\text{a},n}$, and the resulting vorticity and potential-vorticity equations have forcing terms that are identical to the consequences of the Ekman pumping, w_{ek}.

The first conception is the more fundamentally justifiable one, but the second one is generally simpler to use: once $\mathbf{F}$ is specified, the Ekman boundary layers can be disregarded. This equivalence justifies a posteriori the model formulation for the equilibrium zonal baroclinic jet (Section 5.3.1), and it is used again for the equilibrium wind-gyre problem (Section 6.2.4).

6.2.2 Interior and boundary-layer circulations

The steady, linear, barotropic gyre model (6.60) is now solved for a rectangular, flat-bottomed domain and an idealized wind-stress pattern such as that in Fig. 6.12, representing mid-latitude westerly surface winds and tropical and polar easterlies, namely,

$$\tau_s^x(y) = \tau_0 \, \cos\left[\frac{2\pi y}{L_y}\right]$$

$$\implies \frac{\partial \tau_s^x}{\partial y} = -\frac{2\pi\tau_0}{L_y} \, \sin\left[\frac{2\pi y}{L_y}\right]. \tag{6.61}$$

The origin for y is in the middle of the domain in Fig. 6.12. This wind pattern will be shown below to give rise to a *double gyre* pattern of oceanic circulation, with a cyclonic circulation to the north and an anticyclonic one to the south.

Equation (6.60) can be recast in terms of the more familiar velocity stream-function, $\psi = \Psi/H$:

$$D\nabla^2\psi + \frac{\partial\psi}{\partial x} = A \sin\left[\frac{2\pi y}{L_y}\right] \tag{6.62}$$

for

$$D = \frac{f_0}{\beta}\sqrt{\frac{E}{2}} = \frac{\epsilon_{bot}}{\beta H} \quad \text{and} \quad A = \frac{2\pi\tau_0}{\rho_o\beta HL_y}. \tag{6.63}$$

D has the dimensions of length, and A has the dimensions of velocity.

Equation (6.62) is a second-order, elliptic, two-dimensional, partial differential equation in (x, y). It requires one lateral boundary condition on ψ for well-posedness. It comes from the kinematic condition of no normal flow through the boundary,

$$\mathbf{u} \cdot \hat{\mathbf{n}} = \frac{\partial\psi}{\partial s} = 0$$

$$\implies \psi = C \to 0, \tag{6.64}$$

where C is a constant along the boundary. (n, s) are the normal and tangential coordinates at the boundary, located here at $x = 0$, L_x and $y = -L_y/2$, $L_y/2$. For barotropic flow in a simply connected domain, C can be chosen to be zero without loss of generality, since only horizontal gradients of ψ have a physical meaning in this context. (This choice is generally not allowed for shallow-water or baroclinic dynamics since ψ or its purely vertical derivatives appear in the governing equations; cf. (4.113) or (5.28).)

An exact, albeit complicated, analytic solution expression can be written for (6.62)–(6.64). However, it is more informative to find an approximate solution using the *method of boundary-layer approximation* based upon $D \ll L_x$ that is a consequence of $E \ll 1$ by (6.44) and (6.63). Implicitly this method is used in Section 6.1 for the Ekman-layer problem by neglecting horizontal derivatives of the mean fields and treating the finite fluid depth as infinite. In both cases the approximate boundary-layer equations have higher-order derivatives in the boundary-normal direction than arise in either the interior problem or in the along-boundary directions. This method of asymptotic analysis is sometimes called *singular perturbation analysis*.

By neglecting the term of $\mathcal{O}(D)$ in (6.62), the partial differential equation is

$$\frac{\partial \psi}{\partial x} = A \, \sin\left[\frac{2\pi y}{L_y}\right]. \tag{6.65}$$

Note that the highest-order spatial derivatives have disappeared by using this approximation, which will be shown to be appropriate for the interior region but not the lateral boundary layer. Equation (6.65) can be integrated in x, using the boundary condition (6.64) at $x = L_x$:

$$\psi = \psi^{\mathrm{i}}(x, y) = -A \, (L_x - x) \, \sin\left[\frac{2\pi y}{L_y}\right]. \tag{6.66}$$

This expression for the horizontal gyre circulation is called the *Sverdrup transport*. After further multiplication of ψ by H, Ψ is the volume transport around the gyre (NB, but not a column transport; Section 6.1.1).

From (6.61) and (6.66), the Sverdrup transport is

$$\Psi(x, y) = -\frac{1}{\rho_o \beta} \int_x^{L_x} dx \left[\hat{\mathbf{z}} \cdot \mathbf{V}_{\mathrm{h}} \times \boldsymbol{\tau}_{\mathrm{s}}\right].$$

In the interior of the ocean the barotropic streamfunction is proportional to the curl of the wind stress whenever other terms (e.g., bottom drag) in the depth-integrated vorticity balance are negligible, as they are assumed to be here. The origin of this relationship is the formula for the surface Ekman pumping (6.40).

Note that (6.66) satisfies (6.64) at all of the boundaries except the western one, $x = 0$, where

$$\psi^{\mathrm{i}}(0) = -AL_x \, \sin\left[\frac{2\pi y}{L_y}\right] \neq 0. \tag{6.67}$$

The fact that (6.66) satisfies the boundary conditions at $y = -L_y/2$, $L_y/2$ is due to the artful coincidence of the boundary locations with minima in τ_s^x; otherwise, the problem solution would be somewhat more complicated, although essentially similar.

To complete the solution for (6.62)–(6.64), the boundary-layer approximation is made near $x = 0$. Define a non-dimensional coordinate,

$$\xi = \frac{x}{D}, \tag{6.68}$$

and assume that the solution form is

$$\psi(x, y) = \psi^{\mathrm{i}}(x, y) + \psi^{\mathrm{b}}(\xi, y). \tag{6.69}$$

(cf., the Ekman-layer decomposition in (6.16)). Equation (6.68) is substituted into (6.62), and whenever an x-derivative is required for a boundary-layer quantity, it is evaluated with the relation,

$$\frac{\partial}{\partial x} = \frac{1}{D} \frac{\partial}{\partial \xi}.$$

Grouping the terms in powers of D,

$$D^{-1} \left[\frac{\partial^2 \psi^b}{\partial \xi^2} + \frac{\partial \psi^b}{\partial \xi} \right]$$

$$+ D^0 \left[\frac{\partial \psi^i}{\partial x} - A \, \sin \left[\frac{2\pi y}{L_y} \right] \right]$$

$$+ D^1 \left[\nabla^2 \psi^i + \frac{\partial^2 \psi^b}{\partial y^2} \right] = 0. \tag{6.70}$$

By treating D as a small asymptotic ordering parameter (i.e., small compared to L_x), terms of $\mathcal{O}(D^1)$ are negligible. The terms of $\mathcal{O}(D^0)$ cancel by (6.66). So focus on the leading-order terms of $\mathcal{O}(D^{-1})$ to pose the approximate boundary-layer equation for $\psi^b(\xi, y)$:

$$\frac{\partial^2 \psi^b}{\partial \xi^2} + \frac{\partial \psi^b}{\partial \xi} = 0$$

$$\psi^b = -\psi^i(0) \text{ at } \xi = 0$$

$$\psi^b \to 0 \text{ as } \xi \to \infty. \tag{6.71}$$

The first boundary condition assures that there is no normal flow (6.64), and the second condition assures that ψ^b is confined to near the western boundary. The first condition provides the only "forcing" for ψ^b that precludes a trivial solution. The *raison d'être* for the western boundary layer is to divert the interior normal flow at $x = 0$ to be parallel to the boundary; this is equivalent to saying the boundary layer provides a compensating meridional volume transport for the interior Sverdrup transport.

The solution to (6.71) is

$$\psi^b = -\psi^i(0) \, e^{-\xi}, \tag{6.72}$$

and the total solution to (6.62)–(6.64) is approximately

$$\psi(x, y) = -AL_x \, \sin \left[\frac{2\pi y}{L_y} \right] \left(1 - \frac{x}{L_x} - e^{-x/D} \right) \tag{6.73}$$

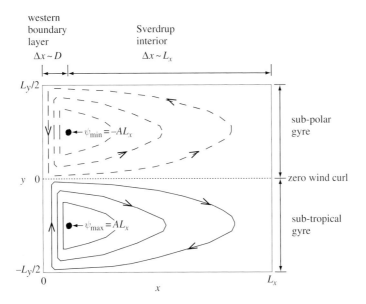

Fig. 6.14. Transport streamfunction, $\Psi(x, y)$, for linear oceanic wind gyres. The sub-polar gyre is to the north (in the northern hemisphere) (indicated by dashed contours for $\Psi < 0$), and the sub-tropical gyre is to the south (indicated by solid contours for $\Psi > 0$).

when $D/L_x \ll 1$. The spatial pattern for ψ is shown in Fig. 6.14. It is composed of two recirculating gyres (i.e., with closed contours of ψ): the *sub-polar gyre* in the north is cyclonic, and the *sub-tropical gyre* in the south is anticyclonic. The gyres are separated by the zero in the wind curl coinciding with the maximum of the westerly winds at $y = 0$. Each gyre has a relatively narrow (i.e., $\Delta x \sim D$) boundary current that connects the interior Sverdrup streamlines with streamlines parallel to the western boundary.

An obvious question is why the boundary current occurs on the western boundary. The answer is that the boundary-layer equation (6.71) has one solution that decays toward the east for $D > 0$ (due to the positive signs of β and the bottom drag, ϵ_{bot}) and none that decays toward the west. The only way that the interior mass transport can be balanced by a boundary-layer transport is for the boundary layer to be on the western side of the basin. This is why it is correct to integrate the Sverdrup solution (6.66) from the eastern boundary with the boundary condition, $\psi^{\text{i}} = 0$, using the single-sided integration constant available for the first-order differential equation (6.65).

The volume transport across any horizontal section is expressed as

$$T_{\perp} = \int_{s_1}^{s_u} ds \int_0^H dz \, \mathbf{u} \cdot \hat{\mathbf{n}} = H \left[\psi(s_u) - \psi(s_1) \right]. \tag{6.74}$$

Here n is the horizontal coordinate across the section, s is the horizontal coordinate along it, and the section spans $s_l \le s \le s_u$. The particular value, $T_\perp = HAL_x$, is the maximum transport in each of the gyres in (6.73) and Fig. 6.14. It represents the transport magnitude between any boundary and the gyre centers just interiorward of the western boundary layer at $y = -L_y/4$ and $L_y/4$.

With this approximate boundary-layer solution, the meridional velocity in (6.73) is

$$v(x, y) = \frac{\partial \psi}{\partial x} \approx -\frac{AL_x}{D} \, \sin\left[\frac{2\pi y}{L_y}\right] \left(e^{-x/D} - \frac{D}{L_x}\right). \qquad (6.75)$$

Its structure is sketched in Fig. 6.15. There is a narrow meridional western boundary current that has a much stronger velocity than the interior Sverdrup flow. It is northward in the sub-tropical gyre and southward in the sub-polar gyre. The meridional volume transports (i.e., the x, z integrals of v) of the boundary-layer and Sverdrup circulations are in balance at every latitude.

6.2.3 Application to real gyres

The western boundary currents in the preceding section can be identified – at least qualitatively with respect to location and flow direction – with the strong, persistent, sub-tropical and sub-polar western boundary currents in the North Atlantic Ocean (the Gulf Stream and the Labrador Current) and North Pacific Ocean

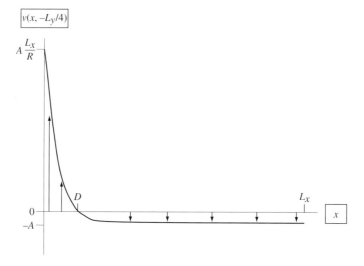

Fig. 6.15. Meridional velocity profile, $v(x)$, for a barotropic oceanic wind gyre across a zonal section that runs through the middle of the sub-tropical gyre (at $y = -L_y/4$). Note the narrow, poleward western boundary current and the broad, equatorward Sverdrup flow in the interior.

(the Kuroshio and the Oyashio Currents). They have sub-tropical counterparts in the southern hemisphere (e.g., the Brazil Current in the South Atlantic, the East Australia Current in the South Pacific, and the Agulhas Current in the South Indian) but no sub-polar ones since the ACC region is zonally unbounded by continents and does not have a wind-gyre circulation (Section 5.3).

How accurate and dynamically consistent is the solution in Section 6.2.2 for real wind gyres? To test its underlying approximations – and to give it a physical interpretation – empirical estimates are made for various properties of the gyre circulation:

$$H = 5 \times 10^3\,\text{m}, \quad L_x \approx L_y = 6 \times 10^6\,\text{m},$$

$$f_0 = 10^{-4}\,\text{s}^{-1}\ (\text{at }45°\,\text{N}), \quad \beta = 2 \times 10^{-11}\,\text{m}^{-1}\,\text{s}^{-1},$$

$$\tau_0 = 0.1\,\text{N m}^{-2} = \rho_o \times 10^{-4}\,\text{m}^2\,\text{s}^{-2},\ h_{\text{ek, top}} = 100\,\text{m},$$

$$\implies v^{\text{i}} \sim A = 10^{-3}\,\text{m s}^{-1}.$$

$$\implies \nu_{\text{e}} = 0.5\,\text{m}^2\,\text{s}^{-1},\ v_{\text{ek, top}} = 10^{-2}\,\text{m s}^{-1},\ \text{and}\ E = 10^{-4}.$$

$$\implies \Delta x^{\text{b}} \sim D = 5 \times 10^4\,\text{m and}\ \frac{D}{L_x} = 0.008.$$

$$\implies v^{\text{b}} \sim \frac{AL_x}{D} = 0.13\,\text{m s}^{-1}.$$

$$\implies \max[T_\perp] = HAL_x = 31.5 \times 10^6\,\text{m}^3\,\text{s}^{-1} = 31.5\,\text{Sv}. \tag{6.76}$$

The magnitudes in the first three lines are chosen from measurements of wind gyres and their environment, and the magnitudes in the final five lines are deduced from the analytic gyre solution in Section 6.2.2. In the third line the unit for force is a *Newton*, $1\,\text{N} = \text{kg m s}^{-2}$. In the final line the unit, $1\,\text{Sv} = 10^6\,\text{m}^3\,\text{s}^{-1}$, is introduced. It is called a *Sverdrup*, and it is the most commonly used unit for oceanic volume transport.

The a-posteriori consistency of the simple gyre model and its solution (Section 6.2.2) can now be checked for several assumptions that were made in its derivation. This kind of analysis is a necessary step to decide whether an approximate GFD analysis is both dynamically self-consistent and consistent with nature.

$E \ll 1$ *and* $D/L_x \ll 1$ From (6.76) these conditions are well satisfied.

$Ro \ll 1$ Estimates from (6.76) give $Ro^{\text{i}} = 2 \times 10^{-6}$ and $Ro^{\text{b}} = 2 \times 10^{-2}$. Both of these Rossby numbers are small, consistent with the assumptions made in deriving the model (6.58).

$|\zeta_t| \ll |\beta v|$ This can be alternatively expressed as a restriction on the time scale of wind variation, t_*, in order for the steady-state response assumption to be valid; namely, in the interior,

$$t_* \gg \frac{\zeta}{\beta v} \approx \frac{1}{\beta L_x} = 10^4 \, \text{s} < 1 \, \text{d}.$$

In the western boundary layer the analogous condition is $t_* \gg (D\beta)^{-1} \approx 10^6 \, \text{s} \approx$ 10 d, which is much more restrictive since $D \ll L_x$. These time scales relate to barotropic Rossby-wave propagation times across the basin and boundary layer, respectively, assuming that the wave scale is equal to the steady current scale. Obviously, one must accept the more stringent of the two conditions and conclude that this theory is only valid for steady or low-frequency wind patterns, rather than passing storms. In the real ocean the mean gyres are baroclinic (i.e., with **u** largely confined in and above the pycnocline), so the relevant Rossby-wave propagation times are baroclinic ones, $L_x/(\beta R_1^2)$ and $D/(\beta R_1^2)$, respectively, with $R_1 = 5 \times 10^4$ m; the former condition is the longer time of ≈ 2 yr. So the steady gyre assumption is only valid for wind fields averaged over several years.

$|\mathbf{u} \cdot \nabla \zeta| \ll |\beta v|$ In the interior this condition is re-expressed as

$$1 \gg \frac{v^{\text{i}}}{\beta L_y^2} \approx 10^{-6},$$

and it is well satisfied (cf., (4.124)). In the boundary layer there is a larger velocity, v^{b}, a shorter cross-shore scale, D, and the same along-shore scale, L_y; so the condition for neglecting advection is

$$\beta v^{\text{b}} \gg v^{\text{b}} \frac{\partial \zeta^{\text{b}}}{\partial y},$$

$$\text{or} \quad 1 \gg \frac{v^{\text{b}}}{\beta D L_y} \approx 2 \times 10^{-2}. \tag{6.77}$$

This seems to be self-consistent. However, it has been discovered by solving computationally for nonlinear, time-dependent solutions of (6.58) that they develop a meridionally narrow region of *boundary-current separation*, unlike the broad separation region for the linear solution in Fig. 6.14 (and the western boundary currents in nature also have a narrow separation region). A sketch of a hypothetical alternative separation flow pattern is shown in Fig. 6.16. If D is

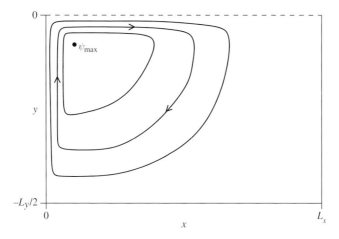

Fig. 6.16. Sketch of the transport streamfunction, $\Psi(x, y)$, for a steady, nonlinear, barotropic sub-tropical gyre in the northern hemisphere. In comparison with the linear solution in Fig. 6.14, note the migration of the gyre center to the northwest and the associated narrow separation of the western boundary current into the interior near the northern gyre boundary.

used as an estimate of both the boundary-layer and separation-flow widths, a more appropriate check on neglecting the nonlinearity is

$$1 \gg \frac{v^b}{\beta D^2} = \frac{L_\beta^2}{D^2} \approx 2, \tag{6.78}$$

which is not well satisfied. In the nonlinear solution in Section 6.2.4 the boundary current width is approximately $L_\beta = (v^b/\beta)^{1/2}$. ($L_\beta$ has the same definition as the Rhines scale (Section 4.8.1), although it has a different meaning in the present context.) Furthermore, if one takes into account the fact that real wind gyres are baroclinic, then v^b is even larger by a factor of approximately $H/h_{\text{pycnocline}}$ to achieve the same boundary-layer transport, and the revised condition (6.78) is even more strongly violated.

   Therefore, it must be concluded that this linear gyre solution is not a realistic one, principally because of its neglect of advection in the western boundary current, even though it is an attractive solution because it has certain qualitative features in common with observed oceanic gyres. (The wind stress would have to be an order of magnitude smaller for the gyre circulation to have self-consistent linear dynamics.) Nonlinear dynamical influences are needed for realism, at least for the western boundary layer and its separated extension into the interior circulation. However, nonlinear mean gyre circulations are usually unstable, as is true for most elements of the oceanic and atmospheric general circulation. This

implies that a truly relevant solution will include transient currents as well as
mean currents and so be yet another example of eddy–mean interaction. Since
the primary instability mechanism is a mixture of barotropic and baroclinic types
(Sections 3.3 and 5.2), a relevant oceanic model must also be fully baroclinic. For
all of these reasons, the model of a linear, steady, barotropic gyre is an instructive
GFD example, but it cannot easily be extended to real oceanic gyres without
major modifications. To demonstrate how a more general model compares with
the idealized analytic model, a computational solution for a baroclinic wind gyre
is examined in the next section.

6.2.4 Turbulent baroclinic wind gyres

The N-layer, adiabatic, quasigeostrophic model (Section 5.2) is also appropriate
for examining the problem of a turbulent statistical equilibrium of the baroclinic
double wind gyre. The problem posed here is nearly the same as for the zonal jet
(Section 5.3), except the side boundaries are closed zonally as well as meridionally,
and the steady surface wind pattern for $\tau_s^x(y)$ (Fig. 6.12) spans a greater meridional
range to encompass both a sub-tropical gyre and a sub-polar gyre (Fig. 6.14).
As in Section 5.3, the gyre solution is based on the β-plane approximation
and a background stratification, $\mathcal{N}(z)$, with a shallow pycnocline (around 600 m
depth) and baroclinic deformation radii, R_m, $m \geq 1$, much smaller than the basin
dimensions, $L_x = 3600$ km and $L_y = 2800$ km (i.e., somewhat smaller than the
value in (6.76) to reduce the size of the computation). The horizontal domain is
rectangular, and the bottom is flat with $H = 5$ km. The vertical layer number is
$N = 8$, which is rather higher than is necessary to obtain qualitatively apt solution
behaviors.

   After spin up from a stratified state of rest over a period of about ten years,
a turbulent equilibrium state is established. The instantaneous flow (Fig. 6.17)
shows the expected sub-tropical ($\psi > 0$) and sub-polar ($\psi < 0$) gyres with narrow
western boundary currents that separate and meander as a narrow jet in the region
between the gyres. These gyre-scale flow patterns are most evident in the upper
ocean, although there is abundant eddy variability as well. At greater depths the
mesoscale eddies are increasingly the dominant type of current. The horizontal
scales of the eddies, boundary current, and extension jet are all comparable
both to the largest baroclinic deformation radius, $R_1 \approx 50$ km, and to the inertial
boundary current scale, $L_\beta = \sqrt{V/\beta} \approx 100$ km based on the boundary current
velocity (6.78) (rather than on the linear bottom-drag scale, D, in Section 6.2.2).
The comparability of these different horizontal scales indicates that the eddies
arise primarily from the instability of the strong boundary current and separated
jet through a mixture of barotropic and baroclinic types because of the presence

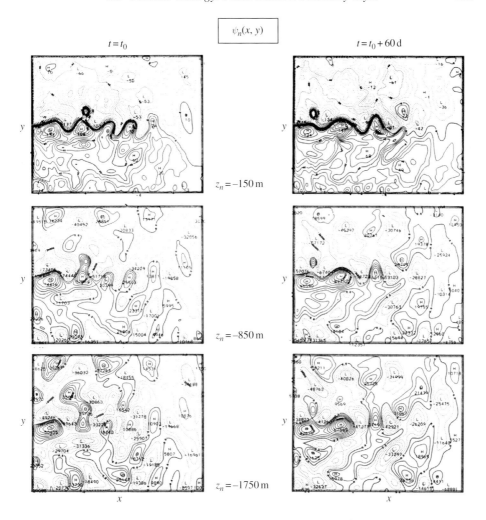

Fig. 6.17. Instantaneous quasigeostrophic streamfunction, $\psi_n(x, y)$ in three differ-ent layers with mean depths of 150, 850, and 1750 m (in rows) from an eight-layer model of a double wind gyre at two different times 60 d apart (in columns). Note the meandering separated boundary current and the break-off of an anticyclonic eddy into the sub-polar gyre (Holland, 1986).

of significant horizontal and vertical mean shears in the boundary currents and elsewhere.

The time-mean flow (Fig. 6.18) does not show the transient mesoscale eddies because of the averaging. The Sverdrup gyre circulation is evident (cf. Fig. 6.14), primarily in the upper ocean and away from the strong boundary and separated jet currents. At greater depth and in the neighborhood of these strong currents are *recirculation gyres* whose peak transport is several times larger than the Sverdrup transport. These recirculation gyres arise in response to downward

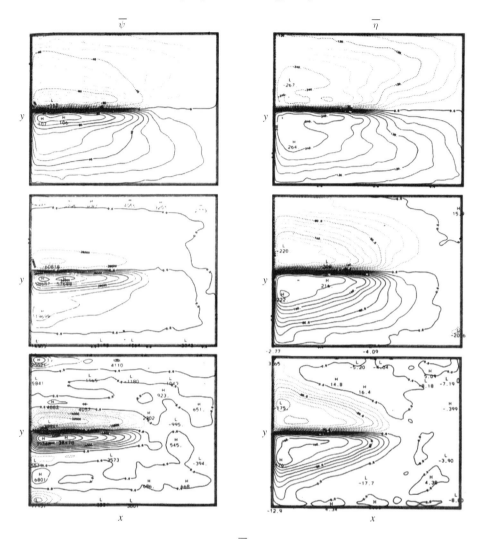

Fig. 6.18. Time-mean streamfunction, $\overline{\psi}_n(x, y)$, at the same mean depths as in Fig. 6.17 (left column) and interface displacement, $\overline{\eta}_{n+0.5}(x, y)$, at mean depths of 300, 1050, and 2000 m (right column) in an eight-layer model of a double wind gyre. Note the Sverdrup gyre in the upper ocean, the separated western boundary currents, and the recirculation gyres in the abyss (Holland, 1986).

eddy momentum flux by isopycnal form stress (cf. Section 5.3.3). The separated time-mean jet is a strong, narrow, surface-intensified current (Fig. 6.19, top). Its instantaneous structure is vigorously meandering and has an eddy kinetic energy envelope that extends widely in both the horizontal and vertical directions away from the mean current that generates the variability (Fig. 6.19, bottom).

Overall, the mean currents and eddy fluxes and their mean dynamical balances have a much more complex spatial structure in turbulent wind gyres than in zonal

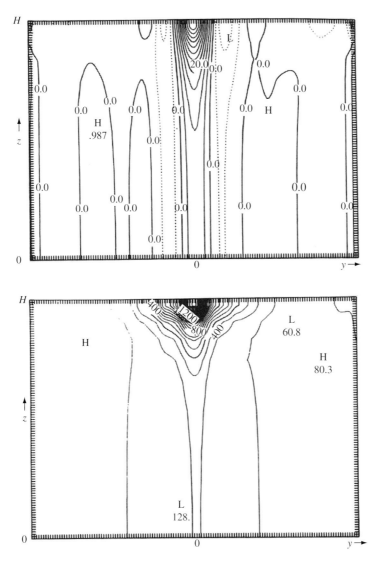

Fig. 6.19. A meridional cross-section at mid-longitude in the basin in an eight-layer model of a double wind gyre. (Top) mean zonal velocity, $\bar{u}$ (y, z). The contour interval is $0.05\,\text{m s}^{-1}$, showing a surface jet maximum of $0.55\,\text{m s}^{-1}$, a deep eastward flow of $0.08\,\text{m s}^{-1}$, and a deep westward recirculation-gyre flow $0.06\,\text{m s}^{-1}$. (Bottom) eddy kinetic energy, $\frac{1}{2}\overline{(\mathbf{u}')^2}$ (y, z). The contour interval is $10^{-2}\,\text{m}^2\,\text{s}^{-2}$, showing a surface maximum of $0.3\,\text{m}^2\,\text{s}^{-2}$ and a deep maximum of $0.02\,\text{m}^2\,\text{s}^{-2}$ (Holland, 1986).

jets. While simple analytic models of steady linear gyre circulations (Section 6.2.2) and their normal-mode instabilities (e.g., as in Sections 3.3 and 6.2) provide a partial framework for interpreting the turbulent equilibrium dynamics, obviously they do so in a mathematically and physically incomplete way. The eddy–mean interaction includes some familiar features – e.g., Rossby waves and vortices; barotropic and baroclinic instabilities; turbulent cascades of energy and enstrophy and dissipation (Section 3.7); turbulent parcel dispersion (Section 3.5); lateral Reynolds stress and eddy heat flux (Sections 3.4 and 5.2.3); downward momentum and vorticity flux by isopycnal form stress (and important topographic form stress if $B \neq 0$) (Section 5.3.3); top and bottom planetary boundary layers (Section 6.1); and regions of potential vorticity homogenization (Section 5.3.4) – but their comprehensive synthesis remains illusive.

Rather than pursue this problem further, this seems an appropriate point to end this introduction to GFD – contemplating the relationship between simple, idealized analyses and the actual complexity of geophysical flows evident in measurements and computational simulations.

Afterword

This book, from necessity, gives only a taster's sampling of the body of posed and partially solved problems that comprises GFD. A comparison of the material here with other survey books (e.g., those cited in the References) indicates the great breadth of the subject, and, of course, the bulk of the scientific record for GFD is to be found in journal articles, only lightly cited here.

Most of the important GFD problems have been revisited frequently. The relevant physical ingredients – fluid dynamics, material properties, gravity, planetary rotation, and radiation – are few and easily stated, but the phenomena that can result from their various combinations are many. Much of the GFD literature is an exploration of different combinations of the basic ingredients, always with the goal of discovering better paradigms for understanding the outcome of experiments, observations, and computational simulations.

Mastery of this literature is a necessary part of a research career in GFD, but few practitioners choose to read the literature systematically. Instead the more common approach is to address a succession of specific research problems, learning the specifically relevant literature in the process. My hope is that the material covered in this book will provide novice researchers with enough of an introduction, orientation, and motivation to go forth and multiply.

Exercises

Fundamental dynamics

1. Consider a two-dimensional (2D) velocity field that is purely rotational (i.e., it has a vertical component of vorticity, $\zeta = \nabla \times \mathbf{u}$, but its divergence, $\delta = \nabla \cdot \mathbf{u}$, is zero in both 2D and 3D):

$$u = -\frac{\partial \psi}{\partial y}, \quad v = +\frac{\partial \psi}{\partial x}, \quad w = 0,$$

where the streamfunction, ψ, is a scalar function of (x, y, t).

(a) Show that an isoline of ψ is a streamline (i.e., a line tangent to $\mathbf{u}$ everywhere at a fixed time).

(b) For the Eulerian expression,

$$\psi = -Uy + A \, \sin[k(x - ct)],$$

sketch the streamlines at $t = 0$. Without loss of generality assume that U, A, k, and c are positive constants. (Note: if the second component of ψ represents a wave, then A is its amplitude; k is its wavenumber; and c is its phase speed.)

(c) For ψ in (b) find the equation for the trajectory (i.e., the spatial curve traced over time by a parcel moving with the velocity) that passes through the origin at $t = 0$.

(d) For ψ in (b) sketch the trajectories for $c = -U$, 0, U, and $3U$; summarize in words the dependence on c. (Hint: pay special attention to the case $c = U$ and derive it as the limit $c \to U$.)

(e) For ψ in (b) show that the divergence is zero and evaluate the vorticity. [Sections 2.1.1 and 2.1.5]

2. T is 3 K cooler 25 km to the north of a particular point, and the wind is northerly (i.e., from the north) at 10 m s^{-1}. The air is being heated by the net of absorbed and emitted radiation at a rate of 0.5 m^2 s^{-3}. What is the local rate of change of temperature? (Assume the motion is along an isobar, so that there is no change of pressure following the flow.) [Section 2.1.2]

3. Derive the vertical profiles of pressure and density for a resting, conservative ocean with temperature and salinity profiles,

$$T(z) = T_0 + T_* e^{z/H}, \quad S(z) = S_0 + S_* e^{z/H},$$

for $z \leq 0$, assuming a Boussinesq equation of state as an approximation to the true equation of state for seawater (i.e., completely neglect compressibility, including the small difference between temperature and potential temperature). Estimate approximate values for T_0, S_0, T_*, S_*, and H in Earth's ocean. [Sections 2.2.1 and 2.3.2]

4. If an atmospheric pressure fluctuation of -10^3 Pa passes slowly over the ocean, how much will the local sea level change? If a geostrophic surface current in the ocean is 0.1 m s^{-1} in magnitude and 100 km in width, how big is the sea-level change across it? Explain how each of these behaviors is or is not consistent with the rigid-lid approximation for oceanic dynamics. [Sections 2.2.3 and 2.4.2]

5. Starting with the equations of state, internal energy, and entropy for an ideal gas, derive the evolution equation for potential temperature (2.52). Estimate the temperature change that occurs if an air parcel is lifted over a mountain 2 km high. [Sections 2.3.1 and 2.3.2]

6. Describe the propagation of sound in the $\hat{\mathbf{x}}$ direction for an initial wave form that is sinusoidal with wavenumber, k, and pressure amplitude, p_*. Derive the relations among p, ρ, $\mathbf{u}$, ζ, and δ. (Hint: linearize the compressible, conservative fluid equations for an ideal gas around a thermodynamically uniform state of rest, neglecting the gravitational force.) [Section 2.3.1]

7. Derive a solution for an oceanic surface gravity wave with a surface elevation of the form

$$h(x, y, t) = h_0 \sin[kx - \omega t], \quad h_0, k, \omega > 0.$$

Assume conservative fluid dynamics, constant density, constant pressure at the free surface, incompressibility, small h_0 (such that the free surface condition can be linearized about $z = 0$ in a Taylor series expansion and that the dynamical equations can be linearized about a state of rest), irrotational motion (i.e., $\zeta = 0$), and infinite water depth (i.e., with all motion vanishing as $z \rightarrow -\infty$). Explain what happens to this solution when the rigid-lid approximation is made. [Section 2.2.3 and Stern (Chapter 1, 1975)]

8. Derive the vertical profiles of pressure and density for a resting atmosphere with a temperature profile,

$$T(z) = T_0 + T_* e^{-z/H},$$

for $z \geq 0$, assuming an ideal gas law. For positive T_0, T_*, and H, what is a necessary condition for this to be a gravitationally stable profile? (Hint: what is required for θ to have a positive gradient?) [Section 2.3.2]

9. Derive the equations for hydrostatic balance and mass conservation in pressure coordinates using alternatively the two different $\tilde{z}$ definitions in (2.74) and (2.75). Discuss comparatively the advantages and disadvantages for these alternative transformations. (Hint: make sure that you use the alternative forms for $\omega \equiv D_t \tilde{z}$, the flow past an isobaric surface.) [Section 2.3.5]

10. Demonstrate the rotational transform relations for D_t, ∇, $\nabla \cdot \mathbf{u}$, and $D_t \mathbf{u}$, starting from the relations defining the rotating coordinates, unit vectors, and velocities. [Section 2.4.1]

11. What are the vertical vorticity and horizontal divergence for a geostrophic velocity for $f = f_0$, and $f = f_0 + \beta_0(y - y_0)$? Explain why $f = f(y)$ is geophysically relevant. In combination with the hydrostatic relation, derive the thermal-wind balance (i.e., eliminate the geopotential between the approximated vertical and horizontal momentum equations). [Section 2.4.2]

12. Redo the scaling analysis in Section 2.3.4 for a rotating flow to derive the condition under which the hydrostatic approximation would be valid for approximately geostrophic motions with $Ro \ll 1$ (i.e., the replacement for (2.72)). Comment on its relevance to large-scale oceanic and atmospheric motions. (Hint: the geopotential,

density, and vertical velocity scaling estimates – expressed in terms of the horizontal velocity and length scales and the environmental parameters – should be consistent with geostrophic and hydrostatic balances.) [Section 2.4.2]

13. Which way do inertial-wave velocity vectors rotate in the southern hemisphere? Which way does the air flow geostrophically around a low-pressure center in the southern hemisphere (e.g., a cyclone in Sydney, Australia)? [Section 2.4.3]

Barotropic and vortex dynamics

1. Show that the area inside a closed material curve – such as the one in Fig. 3.2 – is conserved with time in a 2D flow. (Hint: mimic in 2D the 3D relations in Section 2.1.5 among volume conservation, surface normal flow, and interior divergence.) Does this result depend upon whether or not the non-conservative force, $\mathbf{F}$, is zero? [Section 3.1]

2. Show that the following quantities are *integral invariants* (i.e., they are conserved with time) for conservative 2D flow and integration over the whole plane. Assume that ψ, $\mathbf{u}$, $\zeta \to 0$ as $|\mathbf{x}| \to \infty$, $\mathbf{F} = \mathcal{F} = 0$, and $f(y) = f_0 + \beta_0(y - y_0)$. [Section 3.1]

   (a) $\frac{1}{2} \iint (u^2 + v^2) \, dx \, dy$ [kinetic energy]
      (Hint: multiply the 2D vorticity equation by $-\psi$ and integrate over the area of the whole plane; consider integrations by parts.)
   (b) $\iint \zeta \, dx \, dy$ [circulation]
   (c) $\iint \zeta^2 \, dx \, dy$ [enstrophy]
   (d) $\iint q^2 \, dx \, dy$ [potential enstrophy]
   (e) $\iint x\zeta \, dx \, dy$, $\iint y\zeta \, dx \, dy$ (for $\beta_0 = 0$)   [spatial centroid]

3. For a stationary, axisymmetric vortex with a monotonic pressure anomaly, how do the magnitude and radial scale of the associated circulation in gradient-wind balance differ depending upon the sign of the pressure anomaly when Ro is small, but finite? [Section 3.1.4]

4. Calculate the trajectories of three equal-circulation point vortices initially located at the vertices of an equilateral triangle. [Section 3.2.1]

5. Calculate using point vortices the evolution of a *tripole* vortex with a total circulation of zero. Its initial configuration consists of one vortex between two others that are on opposite sides, each with half the strength and the opposite parity of the central vortex. [Section 3.2.1]

6. Is the weather more or less predictable than the trajectories of N point vortices for large N? Explain your answer. [Section 3.2.2]

7. State and prove Rayleigh's inflection point theorem for an inviscid, steady, barotropic zonal flow on the β-plane. [Section 3.3.1]

8. Derive the limit for the Kelvin–Helmholtz instability of a free shear layer with a piecewise linear profile, $U(y)$, as the width of the layer becomes vanishingly thin and approaches a vortex sheet (i.e., take the limit $kD \to 0$ for the eigenmodes of (3.87)). What are the unstable growth rates, and what are the eigenfunction profiles in y? In particular, what are the discontinuities, if any, in ψ', u', v', and ϕ' across $y = 0$? [Section 3.3.3 and Drazin & Reid (Section 4, 2004). (The latter solves the problem of a vortex-sheet instability for a non-rotating fluid; interestingly, the growth rate formula is the same, but their method, which assumes pressure continuity across the sheet, is not valid when $f \neq 0$.)]

9. Derive eddy–mean interaction equations for enstrophy and potential enstrophy, analogous to the energy equations in Section 3.4. Derive eddy-viscosity relations for

these balances and interpret them in relation to the jet flows depicted in Fig. 3.13. Explain the relationship between eddy Reynolds stress and vorticity flux profiles. [Sections 3.4 and 3.5]

10. Explain how the emergence of coherent vortices in 2D flow is or is not consistent with the general proposition that entropy and/or disorder can only increase with time in isolated dynamical systems. [Sections 3.6 and 3.7]

Rotating shallow-water and wave dynamics

1. Show that the area inside a closed material curve – such as the one in Fig. 3.2 – is not conserved with time in a shallow-water flow. Does this result depend upon whether the non-conservative force, **F**, is zero or not? [Section 4.1]

2. Derive the shallow-water equations appropriate to an active layer between two inert layers with different densities (lighter above and denser below), assuming that the layer interfaces are free surfaces. What is the appropriate formula for a potential vorticity conserved on parcels when $\mathbf{F} = 0$? [Section 4.1]

3. Show that the total energy E defined by

$$E \equiv \iint dx \, dy \, \frac{1}{2} \left(h\mathbf{u}^2 + g\eta^2 \right)$$

is an integral invariant of the conservative shallow-water equations, given favorable lateral boundary conditions. [Section 4.1.1]

4. In conservative shallow-water equations with mean depth H, assume there is a uniform zonal geostrophic flow, $\mathbf{u} = u_0 \hat{\mathbf{x}}$, coming from $x = -\infty$ towards a mountain whose height is ΔH. Calculate the vorticity of fluid parcels that (a) move from $x = -\infty$ onto the top of the mountain, (b) move from on top of the mountain away towards $x = \infty$, and (c) do both movements in sequence. Assume that the surface height variations are smaller than the topographic height variations. (Hint: consider potential vorticity conservation.) [Section 4.1.1]

5. In the conservative shallow-water equations, (a) show that stationary solutions are ones in which $q = \mathcal{G}[\Psi]$, where $\mathcal{G}$ is any functional operator and Ψ is a *transport streamfunction*, such that $h\mathbf{u} = \hat{\mathbf{z}} \times \nabla \Psi$. (b) Show that this implies flow along contours of $f/h = (f + \beta_0 y)/(H - B)$, when the spatial scale of the flow is large enough and/or the velocity is weak enough so that ζ and η are negligible compared to δf and δh. (c) Under these conditions sketch the trajectories for an incident uniform eastward flow across a mid-oceanic ridge with $B = B(x) > 0$ or, alternatively, with $B = B(y) > 0$. What about with steady uniform meridional flow? [Section 4.1.1]

6. (a) Derive the inertia-gravity wave dispersion relation for small-amplitude fluctuations in the conservative 3D Boussinesq equations with a simple equation of state, $\rho = \rho_0(1 - \alpha T)$ (i.e., so that density is conserved on parcels), for a basic state of rest with uniform rotation and stratification, $f = f_0$ and $\overline{\rho}(z) = \rho_0 \left(1 - N_0^2 z/g\right)$, in an unbounded domain. Demonstrate that (b) ω depends only on the direction of **k**, not its magnitude $K = |\mathbf{k}|$, (c) N_0 and f_0 are the largest and smallest frequencies allowed for the inertia-gravity modes, and (d) the phase and group velocities,

$$\mathbf{c}_{\mathrm{p}} \equiv \frac{\omega}{\mathbf{k}} \equiv \omega \mathbf{k}/K^2, \quad \mathbf{c}_{\mathrm{g}} \equiv \frac{\partial \omega}{\partial \mathbf{k}} \equiv \left(\frac{\partial \, \omega}{\partial k^{(x)}}, \frac{\partial \, \omega}{\partial k^{(y)}}, \frac{\partial \, \omega}{\partial k^{(z)}} \right),$$

are orthogonal and have opposite-signed vertical components for the inertia-gravity modes. [Section 4.2 and Holton (Sections 7.4 and 7.5, 2004)]

7. Describe qualitatively the end states of geostrophic adjustment for the following initial sea-level and velocity configurations in the shallow-water equations with $\beta = o$: (a) an axisymmetric mound with no motion and a flat bottom; (b) an axisymmetric depression with no motion and a flat bottom; (c) a velocity patch (i.e., a horizontal square with uniform horizontal velocity inside and zero velocity outside) with a flat surface and bottom; and (d) an initial state with no motion, flat upper surface, and a topographic bump in the bottom. (Hint: consider the gradient-wind balance to compare (a) vs. (b), as well as the two sides across the initial flow direction in (c) for either $Ro = o(1)$ or $Ro = \mathcal{O}(1)$.) [Section 4.3]

8. What controls the time it takes for a gravity wave with small aspect ratio, $H \ll L$, to approach a singularity in the surface shape? What happens in a 3D fluid afterward? [Section 4.4]

9. For a deep-water surface gravity wave propagating in the $\hat{x}$ direction with dispersion relation $\omega^2 = gk$ and orbital velocity components,

$$ u(x, z, t) = \frac{gk\eta_0}{\omega} \, e^{kz} \sin[kx - \omega t] \text{ and } w(x, z, t) = -\frac{gk\eta_0}{\omega} \, e^{kz} \cos[kx - \omega t], $$

derive the Stokes drift velocity profile starting from the Lagrangian trajectory equation (2.1). (Hint: make sure you obtain (4.96) *en route*.) How does it compare with the shallow-water Stokes drift (4.98)? [Section 4.5]

10. Derive the non-dimensional quasigeostrophic potential vorticity equation (4.113) and (4.114) directly from the non-dimensional shallow-water equations (4.109) and (4.110) as $\epsilon \to 0$. (Note: this is an alternative path to the derivation in the text, where the shallow-water potential vorticity equation (4.24) was non-dimensionalized and approximated in the limit $\epsilon \to 0$.) [Section 4.6]

11. Derive the quasigeostrophic potential vorticity equation for a 3D Boussinesq fluid that is uniformly rotating and stratified (i.e., f and N are constant). Assume a simple equation of state, where density is advectively conserved as an expression of internal energy conservation, and assume approximate geostrophic and hydrostatic momentum balances. Assume that Rossby, $Ro = V/fL$, and Froude, $Fr = V/NH$, numbers are comparably small and that density fluctuations, ρ', are small compared to the mean stratification, $\overline{\rho}$; i.e., the dimensional density field is

$$ \rho(x, y, z, t) = \rho_0 \left[1 - \frac{N^2 H}{g} \left(\frac{z}{H} \right) + \frac{Ro}{\mathcal{B}} \, b(x, y, z, t) \right], $$

where the Burger number, $\mathcal{B} = (NH/fL)^2$, is assumed to be $\mathcal{O}(1)$, and $b = -(gH/\rho_0 fVL)\rho'$ is the non-dimensional, fluctuation buoyancy field. Where does the deformation radius appear in this derivation, and what does it indicate about the relative importance of the component terms in q_{QG}? (Hints: use geostrophic scaling to non-dimensionalize the momentum, continuity, and internal energy equations; then form the vertical vorticity equation and use continuity to replace the horizontal divergence (related to the higher-order, ageostrophic velocity field) with the vertical velocity; combine the result with the internal energy equation to eliminate the vertical velocity; then use the geostrophic relations to evaluate the remaining horizontal velocity, vertical vorticity, and buoyancy terms.) [Section 4.6 and Eq. (5.28) in Section 5.1.2]

12. From the shallow-water equations, derive the dispersion relation for a quasigeostrophic topographic wave, assuming a uniform bottom slope upward to the south. Compare it to the Rossby-wave dispersion relation (4.120). [Section 4.7]

13. Solve for the steady-state wave field in the conservative reflection of an incident shallow-water wave impinging upon a western boundary at $x = 0$ for two separate types of waves: (a) an inertia-gravity wave on the f-plane, and (b) a Rossby wave on the β-plane. (Hint: represent the wave field as a sum of incident and reflected components that add up to satisfy the boundary condition of no normal flow.) [Section 4.7]

Baroclinic and jet dynamics

1. Derive the energy conservation law for a conservative two-layer fluid. Also derive it for an N-layer fluid and compare with the two-layer energy conservation law. [Sections 5.1.1 and 5.1.2]
2. Derive the quasigeostrophic approximation to the N-layer energy conservation law in Problem 1. Then decompose the flow into time-mean and time-variable components, and derive the eddy–mean energy conversion terms (referred to as barotropic and baroclinic conversion, respectively). [Sections 3.4 and 5.1]
3. Calculate the vertical modes for $N = 3$ and comment on the relations of the deformation radii and vertical structures, both among the modes and relative to the modes for an $N = 2$ model. (Hint: pose the modal problem for general $g'_{n+0.5}$ and H_n, but then decide whether assuming equal density jumps and layer depths make the solution so much easier that it suffices to illustrate the nature of the answer.) [Section 5.1.3]
4. Evaluate the vertical modes for a two-layer fluid with $H_1 = H_2$. Then evaluate the vertical modes for a continuously stratified fluid with $\mathcal{N}(z) = \mathcal{N}_0$, a positive constant. Compare the two sets of modes. [Section 5.1.3]
5. Derive a Rayleigh theorem for the necessary condition for an inviscid, two-layer, baroclinic instability of a mean zonal flow. Start from

$$\frac{\partial q'_n}{\partial t} + \bar{u}_n \frac{\partial q'_n}{\partial x} + v'_n \frac{\partial \bar{q}_n}{\partial y} = 0$$

with $n = 1, 2$. Then assume a normal-mode form as in (5.56); multiply each layer equation by $-\psi'^*_n$; sum over layers with weights H_n/H (to mimic a volume integral, the horizontal part of which is trivial given this particular normal-mode form); and determine the necessary condition for ω and C to have a non-zero imaginary part. [Section 5.2]
6. Extend and complete (5.77) by deriving an explicit, dimensional expression for the eddy heat flux, $\overline{v'T'}$, in terms of the barotropic modal streamfunction amplitude, $\tilde{\Psi}_0$. [Section 5.2.3]
7. Solve for the marginal stability relation (i.e., where the imaginary part of C is zero) for a two-layer fluid with a mean, baroclinic zonal flow, with $\beta = 0$, but with a linear drag force in each of the layers such that

$$\frac{\partial q'_n}{\partial t} + \bar{u}_n \frac{\partial q'_n}{\partial x} + v'_n \frac{\partial \bar{q}_n}{\partial y} = -\mu \zeta'_n$$

for $\mu > 0$. (Hint: when confronted with complex algebraic relations, be clever rather than brutish in interpreting the answer.) [Section 5.2]
8. Solve the 3D quasigeostrophic baroclinic instability problem for the uniformly rotating and stratified vertical shear flow,

$$\mathbf{u} = Sz\hat{\mathbf{y}}, \quad b = N^2 z + fSx,$$

in a domain that is horizontally unbounded and vertically bounded with $-H/2 \leq z \leq H/2$. Show that the linearized, conservative fluctuation equation is

$$[\partial_t + V\partial_y]q' = 0 \quad \Rightarrow \quad q' = 0,$$

where q' is defined by (5.28) with $\beta = 0$, and the vertical boundary condition is

$$[\partial_t + V\partial_y]\partial_z\psi' - S\partial_y\psi' = 0$$

at $z = \pm H/2$ (cf., Problem 11 in the preceding problem group). Calculate the normal modes with a horizontal wavenumber vector $\mathbf{k} = (k, \ell)$. Compare the growth rate curve, $\sigma(k, \ell)$, with the two-layer problem, and compare the marginally stable wavenumber with the continuously stratified deformation radii in Problem 4 above. Analyze the vertical eigenmode structure as a function of $\sqrt{k^2 + \ell^2}H$. (Note: this is called the Eady problem, and it is an important GFD paradigm for baroclinic instability.) [Sections 5.1.2 and 5.2.1 and Pedlosky (Section 7.7, 1987)]

9. Is there a mean meridional overturning circulation for a two-layer, horizontally homogeneous, vertical shear flow? (Hint: what is $\overline{w}$?) [Sections 5.2 and 5.3.5]

10. How similar are the turbulent, equilibrium zonal jets for different mean forcing: a meridional heating contrast vs. a top-surface zonal stress? [Section 5.3]

11. Starting from the N-layer equations (5.6)–(5.9) and (5.80), derive the mean zonal momentum balance (5.82) for an equilibrium, turbulent, baroclinic zonal flow. Then perform a volume integral and qualitatively describe the roles of the contributing terms. [Section 5.3.3]

12. For a turbulent, quasigeostrophic equilibrium zonal jet, sketch the meridional profiles of the eddy potential vorticity flux (5.91) in each of the layers and discuss how it contributes to the mean potential vorticity balance. Draw vector diagrams of the Eliassen–Palm flux (5.103) in the meridional plane, and interpret the resulting structure in terms of the other dynamical descriptions of equilibrium jet dynamics in Chapter 5. [Sections 5.3.4 and 5.3.7]

13. Make an eddy–mean interaction analysis for the energetics of the Rossby-wave rectification problem. [Sections 3.4 and 5.4]

Boundary-layer and wind-gyre dynamics

1. Consider a well-mixed planetary boundary layer at the top of an ocean with a suddenly imposed surface stress, $\boldsymbol{\tau}_s = \tau_0 \mathcal{H}(t)\hat{\mathbf{x}}$, where $\mathcal{H}(t)$ is the Heaviside step function ($= 0$ for $t < 0$ and $= 1$ for $t > 0$). Assume $\partial_z(\overline{u}, \overline{v}) = 0$ within the layer $0 \geq z \geq -h_0$. Also assume $(\overline{u'w'}, \overline{v'w'}) = 0$ for $z \leq -h$. Solve the time-dependent spin-up problem, assuming a state of rest for $t \leq 0$. (Hint: The answer is a combination of Ekman currents and inertial oscillations.) [Section 6.1]

2. Solve the steady-state, laminar, Ekman layer problem for a uniform-density ocean with $\boldsymbol{\tau}_s = \tau_0\hat{\mathbf{x}}$ at its top surface, including the additional effect of a steady-state surface gravity wave field with mean Lagrangian velocity (i.e., Stokes drift), $\mathbf{u}^{st}(z) = U_0 e^{kz}\hat{\mathbf{x}}$, which contributes an additional "wave-induced vortex force" to the low-frequency (i.e., wave-averaged) boundary-layer horizontal momentum balance,

$$-f(\overline{v} + v^{st}) = \nu_e \partial_{zz}\overline{u}$$

$$+f(\overline{u} + u^{st}) = \nu_e \partial_{zz}\overline{v},$$

but leaves the boundary conditions on **u** unaltered. What is the direction of boundary layer transport, relative to wind and waves? [Sections 4.5 and 6.1]

3. How should the formula (6.49) for $\bar{u}(z)$ in the surface layer be interpreted as $z \to 0$ from within the fluid? [Section 6.1.7]

4. How would stable or unstable density stratification (e.g., imposed by a stablizing or destablizing buoyancy flux through the surface) change the mean velocity profile of a rotating, shear planetary boundary layer above a solid surface? (Hint: give a qualitative answer based on the effect of stratification on the turbulent intensity, hence momentum transport efficiency.) [Section 6.1]

5. Solve an alternative version of the mid-latitude oceanic wind gyre problem, in which damping by bottom drag (via the bottom Ekman layer) is replaced by turbulent horizontal momentum transport (i.e., Reynolds stress) via mesoscale eddies, with an eddy-viscosity parameterization form of

$$\overline{\mathbf{u}'\mathbf{u}'} = -\nu_e^* \nabla \overline{\mathbf{u}} \ .$$

This contribution to **F** leads to a boundary-value problem that replaces (6.62) in the text:

$$-Q\nabla^4 \psi + \psi \partial_x = A \sin \left[\frac{2\pi y}{L_y} \right],$$

where $Q \equiv \nu_e^*/\beta > 0$. Perform a boundary-layer analysis assuming $Q^{1/3} \ll L_x$. Because this alternative problem is fourth-order in its spatial derivatives, it requires two boundary conditions; use the no-slip conditions, $u = v = 0$ at $x = 0, L_x$ and $y = -L_y/2, L_y/2$. Note that these require boundary layers at all boundaries to render the Sverdrup circulation consistent with the boundary conditions, although the western one is again the most important. (Hint: focus primarily on the western boundary, although all of the boundaries have boundary layers in this alternative problem.) [Section 6.2.2 and Munk (1950)]

6. As yet another alternative for the western boundary current, consider an advective dynamics. If vorticity advection is retained in (6.55), then derive the following alternative barotropic boundary layer equation for $\psi^b(\xi, y)$ that neglects eddy-viscous effects:

$$J_{\xi,y} \left[\psi^i + \psi^b, \ \frac{\partial^2 \psi^b}{\partial \xi^2} + \beta y \right] = 0$$

with the same boundary conditions as in (6.71). Here $\xi = x/(\epsilon L_x)$ is a western boundary-layer coordinate for some $\epsilon \ll 1$ to be determined in solving this equation. The general solution of such a Jacobian equation is that one argument is a functional of the other (cf., Section 3.1.4). For simplicity consider a linear functional relation,

$$\frac{\partial^2 \psi^b}{\partial \xi^2} + \beta y = C \left(\psi^i + \psi^b \right) + D,$$

for some constants, C and D, and a local, linear approximation to the Sverdrup circulation near the western boundary, $\psi^i(x \approx 0, y) \approx -U_0(y - y_0)$. Determine C and D from the $\xi \to \infty$ boundary condition, then solve for ψ^b and identify the formula for ϵ. Show that $U_0 > 0$ is a necessary condition for a valid boundary-layer solution, and discuss this constraint in relation to the sub-tropical gyre patterns in Figs. 6.14 and 6.16. Finally, evaluate the value of ϵ in terms of the gyre quantities in (6.76) and comment on the plausibility of a barotropic, advective western boundary layer as a model for the Gulf Stream. Would its plausibility change if the interior circulation were baroclinically confined to the ocean above the pycnocline? [Sections 6.2.2–6.2.4 and Charney (1955)]

7. Assume that a western boundary current is unstable to mesoscale eddies, and make a scale estimate for the amplitude and 3D length scales for the eddies in terms of the scales for the mean wind gyre. Use the eddy scale estimates to justify their likely geostrophic and hydrostatic balances. [Sections 6.2.3 and 6.2.4]

8. For quasigeostrophic, turbulent, equilibrium wind gyres, derive the time-mean balance equation for a circulation integral on a path following a closed streamline for the mean barotropic velocity. (Hint: do a volume integral of the mean vorticity equation within such a streamline.) Interpret the signs of the contributing terms when the streamline is within a sub-tropical gyre. How do their signs change when the streamline is within a sub-polar gyre? [Section 6.2.4]

9. Explain the spatial and dynamical relationships among the recirculation gyre, the Sverdrup transport, and the western boundary current. What is the role of mesoscale eddies in their maintenance? [Sections 5.3.7 and 6.2.4]

References

Angevine, W. M., Grimsdell, A. W., Hartten, L. M., and Delany, A. C. (1998). The Flatland boundary layer experiments. *Bull. Amer. Met. Soc.* **79**, 419–431.

Batchelor, G. (1967). *An Introduction to Fluid Dynamics* (Cambridge: Cambridge University Press).

Bohren, C. and Albrecht, B. A. (1998). *Atmospheric Thermodynamics* (Oxford: Oxford University Press).

Charney, J. G. (1955). The Gulf Stream as an inertial boundary layer. *Proc. Nat. Acad. Sci. USA* **41**, 731–740.

Coleman, G. N. (1999). Similarity statistics from a direct numerical simulation of the neutrally stratified planetary boundary layer. *J. Atmos. Sci.* **56**, 891–900.

Comte, P. (1989). Ph.D. Thesis. National Polytechnic Institute, Grenoble. Also in Lesieur, M. (1995). Mixing layer vortices. In *Fluid Vortices*, ed. S. I. Green (Dordrecht: Kluwer Press) pp. 35–63.

Couder, Y. and Basdevant, C. (1986). Experimental and numerical study of vortex couples in two-dimensional flows. *J. Fluid Mech.* **173**, 225–251.

Cushman-Roisin, B. (1994). *Introduction to Geophysical Fluid Dynamics* (Englewood Cliffs, NJ: Prentice-Hall).

Drazin, P. G. and Reid, W. H. (2004). *Hydrodynamic Stability* (Cambridge: Cambridge University Press).

Durran, D. R. (1993). Is the Coriolis force really responsible for the inertial oscillation? *Bull. Amer. Met. Soc.* **74**, 2179–2184.

Eliassen, A. and Lystad, M. (1977). The Ekman layer of a circular vortex: a numerical and theoretical study. *Geophysica Norwegica* **31**, 1–16.

Fofonoff, N. P. (1962). Dynamics of ocean currents. *Sea* **1**, 3–30.

Gill, A. E. (1982). *Atmosphere–Ocean Dynamics* (New York: Academic Press).

Holland, W. R. (1986). Quasigeostrophic modeling of eddy-resolving ocean circulation. In *Advanced Physical Oceanographic Numerical Modeling*, ed. J. G. O'Brien (Norwell: Reidel Publishing Co.), pp. 203–231.

Holton, J. R. (2004). *An Introduction to Dynamic Meteorology* (4th edn.), (San Diego, CA: Academic Press).

Kalnay, E. *et al.* (1996). The NCEP/NCAR 40-year reanalysis project. *Bull. Amer. Met. Soc.* **77**, 437–471.

Lighthill, M. J. (1978). *Waves in Fluids* (Cambridge: Cambridge University Press).

Lorenz, E. N. (1967). *The Nature and Theory of the General Circulation of the Atmosphere* (Geneva: World Meterological Organization).

Marchesiello, P., McWilliams, J. C., and Shchepetkin, A. (2003). Equilibrium structure and dynamics of the California Current System. *J. Phys. Ocean* **33**, 753–783.

McWilliams, J. C. (1983). On the relevance of two-dimensional turbulence to geophysical fluid motions. *J. Mechanique* **Numero Special**, 83–97.

McWilliams, J. C. (1984). Emergence of isolated coherent vortices in turbulence. *J. Fluid Mech.* **146**, 21–43.

McWilliams, J. C. (1991). Geostrophic vortices. In *Nonlinear Topics in Ocean Physics: Proc. International School of Physics "Enrico Fermi"*, Course 109, ed. A. R. Osborne (Amsterdam: Elsevier), pp. 5–50.

McWilliams, J. C. and Flierl, G. R. (1979). On the evolution of isolated nonlinear vortices. *J. Phys. Ocean.* **9**, 1155–1182.

McWilliams, J. C. and Chow, J. H. S. (1981). Equilibrium geostrophic turbulence: I. A reference solution in a β-plane channel. *J. Phys. Ocean.* **11**, 921–949.

Munk, W. H. (1950). On the wind-driven ocean circulation. *J. Meteorology* **7**, 79–93.

Niiler, P. P., Maximenko, N. A., and McWilliams, J. C. (2003). Dynamically balanced absolute sea level of the global ocean derived from near-surface velocity observations. *Geophys. Res. Lett.* **30**, 7-1–7-4.

Pedlosky, J. (1987). *Geophysical Fluid Dynamics*, (2nd edn.) (Berlin: Springer-Verlag).

Pedlosky, J. (2003). *Waves in the Ocean and Atmosphere. Introduction to Wave Dynamics* (Berlin: Springer-Verlag).

Salmon, R. (1998). *Lectures on Geophysical Fluid Dynamics* (Oxford: Oxford University Press).

Steele, M., Morley, R., and Ermold, W. (2001). PHC: a global ocean hydrography with a high quality Arctic Ocean. *J. Climate* **14**, 2079–2087.

Stern, M. E. (1975). *Ocean Circulation Physics* (New York: Academic Press).

Stommel, H. (1948). The westward intensification of wind-driven ocean currents. *Trans. AGU* **29**, 202–206.

Whitham, G. B. (1999). *Linear and Nonlinear Waves* (New York: John Wiley and Sons).

Index